Transient Signals on Transmission Lines

An Introduction to Non–Ideal Effects and Signal Integrity Issues in Electrical Systems

Transient Signals on Transmission Lines: An Introduction to Non-Ideal Effects and Signal Integrity Issues in Electrical Systems
Andrew F. Peterson and Gregory D. Durgin

ISBN: 978-3-031-00581-7 paperback

ISBN: 978-3-031-01709-4 ebook

DOI: 10.1007/978-3-031-01709-4

A Publication in the Springer series

SYNTHESIS LECTURES ON COMPUTATIONAL ELECTROMAGNETICS #24

Series Editor: Constantine A. Balanis, Arizona State University

Series ISSN
ISSN 1932-1252   print
ISSN 1932-1716   electronic

Transient Signals on Transmission Lines

An Introduction to Non–Ideal Effects and Signal Integrity Issues in Electrical Systems

Andrew F. Peterson and Gregory D. Durgin
Georgia Institute of Technology

SYNTHESIS LECTURES ON COMPUTATIONAL ELECTROMAGENETICS #24

ABSTRACT

This lecture provides an introduction to transmission line effects in the time domain. Fundamentals including time of flight, impedance discontinuities, proper termination schemes, nonlinear and reactive loads, and crosstalk are considered. Required prerequisite knowledge is limited to conventional circuit theory. The material is intended to supplement standard textbooks for use with undergraduate students in electrical engineering or computer engineering. The contents should also be of value to practicing engineers with interests in signal integrity and high-speed digital design.

KEYWORDS

crosstalk, digital design, impedance matching, reflection, signal intergrity, telegrapher's equations

Preface

The material that follows consists of lectures on the topic of transient signals on transmission lines. Emphasis has been placed on aspects of the subject that have application to signal integrity and high-speed digital circuit design issues, including proper termination schemes to avoid impedance discontinuities, reactive and nonlinear loads, and an introduction to crosstalk. This material has formed the first part of the core undergraduate electromagnetic fields course at the Georgia Institute of Technology since 1999. Since transmission line transients have been de-emphasized in most current textbooks, including those that have been used at Georgia Tech during this time, this material was prepared to supplement traditional texts. With the exception of the material on crosstalk, the authors typically cover each chapter that follows in one 50-minute class period.

Acknowledgments

Over the years, the authors have assimilated ideas and perspectives from many of our teachers and colleagues. During the years that this material was being used at Georgia Tech, homework problems and other suggestions originating from others who taught ECE 3025 have been incorporated. We would especially like to acknowledge the contributions, direct and indirect, of I. M. Besieris, P. W. Klock, P. E. Mast, N. N. Rao, W. T. Rhodes, P. G. Steffes, and M. Swaminathan. The first author also acknowledges the influence of S. Rosenstark's 1994 book *Transmission Lines in Computer Engineering*, from which the crosstalk development in Chapter 10 is adapted.

Contents

CHAPTER 1

Introduction

Objectives: Provide motivation by introducing the students to the finite velocity of electrical signals and the remedy afforded by transmission line theory. Derive the transmission line equations using physical reasoning and Kirchhoff's laws.

Greetings! This is a course about electric circuits and *how they really work*. It would be difficult for most of us to imagine life today without the convenience afforded by electric circuits, from household appliances to video entertainment devices to electronic ignitions in automobiles to computers at home and work. Electric circuits have changed in two major respects over the last 50 years: They have shrunk dramatically in size, from the typical chassis housing a vacuum tube circuit (25 cm on a side) to the integrated circuits ubiquitous in modern "toys" (micrometers in size). As their size shrunk by a factor of more than 1000, the electrical signals existing within them have increased in frequency content by a similar factor. Fifty years ago, most consumer goods involved no more than kilohertz frequencies, although FM radio and VHF television involved frequencies on the order of 100 MHz; today, personal communication devices such as cellular telephones involve frequencies exceeding 10 GHz and the clock rates of personal computers exceed several gigahertz. This revolution in electronics was made possible by a property unique to integrated circuits: as transistor size decreases, performance increases, and production cost decreases [1].

Because of these changes, the old approaches (Kirchhoff's laws) for analyzing and designing circuits do not always work: a new approach is needed. Actually, the approach is not really new, but its application to many circuit problems is. The approach we speak of is *transmission line theory*. It is not new — it was developed in the 1800s for telegraph applications, and it has found continuous use since then for modeling circuits that work at high frequencies, such as those within that VHF TV of 50 years ago. It is therefore no surprise that transmission line theory is alive and well for applications involving gigahertz frequencies in wireless personal communication applications. But there is also a new stomping ground for transmission line concepts: high-performance digital circuits [2–3]. As clock rates approach the gigahertz range, digital circuits seldom behave in the manner predicted by models appropriate for lower frequencies. Therefore, there is a broader audience for transmission line theory among today's engineering students, and that audience now includes computer engineers as well as those interested in communications, RF and millimeter-wave circuits, and other applications such as power distribution networks and cable TV (CATV) systems.

Wait a minute, you ask! What do we mean when we say that Kirchhoff's laws do not work? Were you not just required to take an entire course devoted to those laws? Yes, unfortunately, those laws work for linear, lumped parameter circuits — circuits that can be represented with a finite number of elements such as resistors, capacitors, and inductors. In the real world, signals actually propagate through a continuum. Furthermore, the finite speed of electrical signal propagation becomes an important consideration. Electrical signals propagate at the speed of light, which has the value $c = 299,792,458$ m/s in a vacuum,[1] and somewhat slower than that on a typical printed wiring board (PWB). Kirchhoff's laws neglect the finite velocity of an electrical signal and therefore fail when the time delay or phase shift due to that finite velocity becomes significant. This is seldom a concern at lower frequencies found in audio amplifiers, household appliances, and many other devices. However, at higher frequencies the effect can be pronounced.

If we think in the time domain, the critical parameter is the transit time (or *time of flight*) of a signal relative to the size of the circuit. The velocity of 3×10^8 m/s in air is roughly equivalent to a time delay of 1 ns per foot of travel. A time interval of 1 ns seems small but not if the clock rate of the circuit is 1 GHz! Under these conditions, it may take an entire clock period for a signal to travel across a 6-in PWB, and part of the circuit may be a full clock period behind. This time delay is neglected by lumped parameter circuit analysis that treats this pathway as a simple short circuit. Early integrated circuits involved TTL devices with internal delays of 15 ns or more, limiting the clock rates to lower levels and making the transit time of the signals negligible in comparison. Modern devices have improved to the point where transit delays are the limiting factor in digital circuit design. These transit delays are also known as *latency*. Transmission line theory explicitly includes this time delay, and therefore is applicable to these situations.

If we think in terms of the frequency domain, the key parameter is the wavelength of the signal relative to the size of the circuit. For a sinusoidal signal at frequency f, the wavelength λ is given by

$$\lambda = \frac{v}{f}, \qquad (1.1)$$

where v denotes the velocity of the signal in the medium of concern. For an electrical signal, this velocity is that of light. Note that as the frequency increases, the wavelength decreases. At 1 GHz, the wavelength in air is approximately 30 cm. One wavelength corresponds to 360° of phase shift. If the signal accumulates more than a small fraction of 360° of phase shift in traveling across a circuit, lumped parameter circuit analysis no longer accurately describes the situation. Transmission line theory does.

[1]The definition of the meter was adjusted in 1983 so that the speed of light in a vacuum is now an integer number of meters per second.

We should digress for a moment to make a couple of things more precise. Most situations that we will encounter will involve dielectric materials, which are described in terms of a parameter known as the relative permittivity ε_r (also called the dielectric constant). Magnetic materials (most often encountered in power distribution applications) are described by a relative permeability μ_r. The parameters ε_r and μ_r are unitless scale factors. The speed of light, or of any electrical signal, in a general material is given by

$$\text{speed of light (in m s}^{-1}) = \frac{2.998 \times 10^8}{\sqrt{\varepsilon_r \mu_r}} \tag{1.2}$$

This velocity is always slower than that of light in air. Combining Equations (1.1) and (1.2), we see that the wavelength of a single-frequency signal in a medium characterized by ε_r and μ_r is given by

$$\lambda = \frac{1}{\sqrt{\varepsilon_r \mu_r}} \frac{2.998 \times 10^8}{f} \tag{1.3}$$

where f is the frequency (in Hz) and λ is the wavelength (in meters). Table 1.1 provides dielectric constants for common materials. Note that, as suggested by the table heading, the relative permittivity of a material depends on the frequency of the signal of concern. (This also implies that the velocity of propagation depends on the frequency, which contributes to a form of distortion known as *dispersion* for signals that have energy distributed over a wide band of frequencies.)

There are other "nonideal" effects that can also be taken into account using transmission line theory, including reflections and the associated reflection noise, crosstalk between closely spaced traces, simultaneous switching noise due to the inductance in the power supply path connecting active drivers, etc. These topics will be considered in the chapters to follow. This aspect of circuit analysis is often referred to as *signal integrity*, since the desired signal is corrupted by these effects. In digital applications, nonideal effects are usually thought of as additional noise introduced into the system.

A modern electrical device often involves the construction shown in Figure 1.1, consisting of a chip located on a package, which in turn is attached to a PWB. Signals must travel from the PWB through the package to the chip and back again, while power and electrical ground circuitry follows a similar meandering path. In a complex system such as a computer, various PWBs are linked together with signals routed between them using cables of some form or a backplane. In some situations, a driver chip is located on one PWB while the receiver is located on another. The chip is obviously much smaller than the PWB and made from different materials, and therefore limiting factors differ in each of these environments. Furthermore, the chip may be connected to the package and

TABLE 1.1: Relative permittivities of assorted materials (ε_r at approximately 1 GHz)	
Air	1.0
Alumina	9.4
A-35 ceramic	5.6
Glazed ceramic	7.2
Gallium arsenide	13.0
Germanium	16.0
Glass epoxy	4.0
FR4	4.9
Lucite	2.6
Mica	6.0
Nylon	3.5
Plexiglas	2.6
Polyethylene	2.3
Polyimide	3.5
Polystyrene	2.6
Quartz	3.5
Rexolite 1422	2.5
Silicon	11.8
Silicon dioxide	3.9
Teflon	2.1

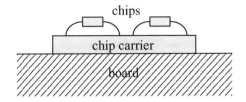

FIGURE 1.1: Construction representative of many modern electrical devices. Chips may include lines and capacitors, as well as active devices like logic gates. The chip carrier contains transmission lines; the board may contain both lines and R, C, and L components.

board in one of several ways, including wirebonding (small individual wires connected between the chip pads and package leads) and ball grid arrays (metallic balls connected directly between the chip pads and the PWB). For accurate analysis under a variety of circumstances that do arise in practice, signals in each of these environments may need to be treated using transmission line theory.

So, what exactly is transmission line theory? To illustrate, consider the two cable cross sections shown in Figure 1.2. This figure shows a coaxial cable (a very popular type of transmission line widely used in microwave measurements and CATV systems) and a parallel strip transmission line (which might approximate interconnects on certain PWBs or chips). When electricity flows on either of these cables, there is a physical movement of charge carriers (electrons) down one conductor and back on the other. This flow of current involves energy stored in the magnetic field associated with the current. This effect is equivalent to some series inductance. There is also some series resistance since the metal is not a "perfect" conductor of electricity, and therefore some of the electrical energy is converted to heat as the current flows. At the same time, equal and opposite charge is stored instantaneously on the two conductors, giving rise to energy stored in the electric field and some shunt capacitance. If the material separating the conductors is not a perfect insulator, there will also be some leakage current from one conductor to another, which we will model as shunt conductance.

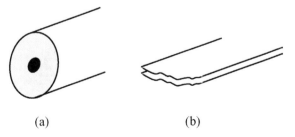

(a)   (b)

FIGURE 1.2: Common transmission lines: (a) coaxial cable and (b) parallel conducting strips.

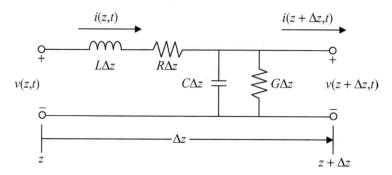

FIGURE 1.3: General model, appropriate for small Δz.

Thus, we can consider an equivalent circuit for a short section of transmission line (say, of length Δz, where Δz is much smaller than a wavelength or the equivalent time of flight) to be that depicted in Figure 1.3. The inductance and capacitance present in this equivalent circuit provide the time delay and phase shift lacking in the treatment of these transmission lines using conventional circuit theory (which would assume that they provide a direct connection from one end to the other). We can therefore analyze this equivalent lumped parameter circuit segment using Kirchhoff's voltage and current laws.

In Figure 1.3, the quantity $L\Delta z$ (H) is the total series inductance of the equivalent circuit, which depends on the inductance per unit length L (H/m). $C\Delta z$ (F) is the total capacitance, which depends on the shunt capacitance per unit length C (F/m). The total series resistance $R\Delta z$ and shunt conductance $G\Delta z$ depend on the resistance per unit length R (Ω/m) and the conductance per unit length G (S/m), respectively.

The application of Kirchhoff's voltage law to the equivalent circuit in Figure 3 yields the equation

$$v(z + \Delta z, t) + L\Delta z\frac{\partial i}{\partial t} + R\Delta z i(z, t) = v(z, t) \tag{1.4}$$

which can be rewritten in the form

$$\frac{v(z + \Delta z, t) - v(z, t)}{\Delta z} = -Ri(z, t) - L\frac{\partial i}{\partial t} \tag{1.5}$$

In the limiting case, as Δz tends to zero, this equation becomes

$$\frac{\partial v}{\partial z} = -Ri(z, t) - L\frac{\partial i}{\partial t} \tag{1.6}$$

An application of Kirchhoff's current law to the circuit in Figure 3 produces

$$i(z + \Delta z, t) - i(z, t) = -G\Delta z v(z + \Delta z, t) - C\Delta z \frac{\partial v}{\partial t} \qquad (1.7)$$

which, in the limiting case as $\Delta z \to 0$, yields

$$\frac{\partial i}{\partial z} = -Gv(z, t) - C\frac{\partial v}{\partial t} \qquad (1.8)$$

Equations (1.6) and (1.8) are known as the *transmission line equations*. In the past, these were called the *telegrapher's equations* in view of the fact that they were originally derived in the 1800s for that application. We note in passing that these equations can also be derived directly from Maxwell's equations (the equations describing electromagnetic fields) applied to a specific transmission line geometry [4–5].

In the following chapter, we will study properties of the solution of the transmission line equations. Before closing, we observe that there are several conventions for sketching transmission lines, and we illustrate those in Figure 1.4.

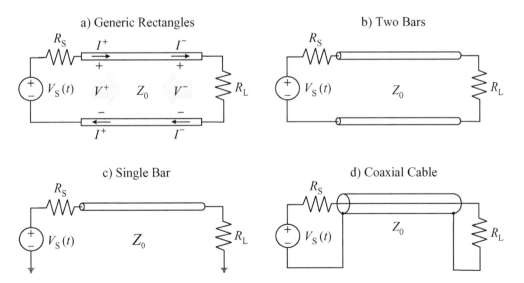

FIGURE 1.4: Common conventions for drawing transmission line systems. A single transmission line is often drawn as two parallel bars or wires, as in (a) and (b). Alternatively, it can be drawn as a single wire with the return path indicated by connections to a common ground (c). In some cases, it is drawn as a coaxial cable (d). In all conventions, the primary line current is assumed to flow from left to right.

Example: A trace runs through a material with relative permittivity $\varepsilon_r = 12$ and relative permeability $\mu_r = 1$. (a) What is the propagation delay through 1 cm of this material? (b) If a sinusoidal signal with frequency 3 GHz is sent through the material, what is the phase shift, in degrees, experienced by the wave through 1 cm of this material?

Solution: (a) The propagation velocity is given by

$$v_p = \frac{c}{\sqrt{\varepsilon_r \mu_r}} \cong \frac{3 \times 10^8}{\sqrt{12}} = 8.654 \times 10^7 \text{ m s}^{-1}$$

Therefore, the propagation delay per meter length of material is given by

$$1.155 \times 10^{-8} \text{ s m}^{-1} = 115.5 \text{ ps cm}^{-1}$$

It follows that the delay through 1 cm is 115.5 ps.

(b) The wavelength is given by

$$\lambda \cong \frac{3 \times 10^8}{\sqrt{12}(3 \times 10^9)} = 0.0289 \text{ m} = 2.89 \text{ cm}$$

Since 1λ corresponds to a phase shift of 360°, a 3-cm distance corresponds to $\dfrac{3}{2.89} \times 360°$ $= 373.7°$ of phase shift.

REFERENCES

[1]   R. D. Isaac, "Reaching the limits of CMOS technology," keynote address, *IEEE Seventh Topical Meeting on the Electrical Performance of Electronic Packaging*, West Point, New York, p. 3, October 1998. doi:10.1109/EPEP.1998.733476

[2]   S. Rosenstark, *Transmission Lines in Computer Engineering*. New York: McGraw-Hill, 1994. doi:10.2200/S00044ED1V01Y200609DCS005

[3]   J. Davis, *High-Speed Digital Design*. Morgan/Claypool Synthesis Lectures on Digital Circuits & Systems #5, 2006.

[4]   N. N. Rao, *Elements of Engineering Electromagnetics*, 2nd edition. Englewood Cliffs, NJ: Prentice-Hall, 1987.

[5]   D. C. Cheng, *Field and Wave Electromagnetics*, 2nd edition. Reading, MA: Addison-Wesley, 1989.

PROBLEMS

1.1 *Fill in the blank:* When the transit time of a transmission line causes delays in pulses that threaten the synchronization of a high-speed digital circuit, we call this effect _____.

1.2 *Fill in the blank:* When transmission line reflections cause data-distorting echoes, we call this effect _____.

1.3 A trace runs through a material with $\varepsilon_r = 4.9$ and $\mu_r = 1$.

    (a) What is the propagation delay through 4 cm of this material?

    (b) If a sinusoidal signal at frequency $f = 100$ MHz is sent through this material, how many degrees of phase shift arise over the 4-cm length?

1.4 As a general rule, transmission line effects become important if a circuit component becomes larger than 0.1λ, where λ is the wavelength of the excitation. Using this rule, decide which of the following situations require transmission line theory:

    (a) A 60-Hz power line spans the 120-mile distance between Atlanta, GA, and Greenville, SC. The velocity of propagation on the line is 8×10^7 m/s.

    (b) A residential telephone line connects a home phone with a public digital network. The distance from the home to the nearest switch is 100 m. The highest major frequency component in transmitted speech is 2 kHz, and the velocity of propagation on the wire is 5×10^7 m/s.

    (c) The radio hardware of a cellular tower is located on the ground and uses long coaxial cables to connect with the antennas at the top of 60 m towers. The operating frequency is 1920 MHz, and the dielectric parameters of the coaxial cable are $\varepsilon_r = 5.3$ and $\mu_r = 1$.

    (d) A memory chip is connected to a microprocessor on a computer motherboard with 4-cm microstrip lines. The velocity of propagation is 1.2×10^8 m/s, and the highest harmonic content of the binary signaling is 3 GHz.

    (e) A 25-ft Ethernet cable transmits binary signals with 100 MHz of maximum frequency content between a router and a computer. The dielectric parameters of the cable are $\varepsilon_r = 2.7$ and $\mu_r = 1$.

1.5 In September 2003, Sun Microsystems announced a new approach for transmitting data within a computer. By placing the edge of one integrated circuit chip directly in contact with another and avoiding the connection through a printed circuit board, they proposed to move data at rates up to 100 times faster than by conventional means.

    Using the principles of transmission line theory, explain why it might be advantageous to connect chips in this fashion.

1.6 List five detrimental effects that can result from propagation on transmission lines.

CHAPTER 2

Solution of the Transmission Line Equations

Objectives: Discuss the general solution of the transmission line equations and explore the meaning of a "one-dimensional wave." Summarize the line parameters for a number of practical transmission lines.

Chapter 1 derived the transmission line equations given in Equations (1.6) and (1.8). For the present, we will neglect the losses due to the noninfinite conductivity of the metal in the line and the nonzero conductivity of the material separating the conductors in order to make the simplifying assumption that $R = 0$ and $G = 0$. Consequently, the transmission line equations for lossless lines reduce to

$$\frac{\partial v}{\partial z} = -L\frac{\partial i}{\partial t} \tag{2.1}$$

$$\frac{\partial i}{\partial z} = -C\frac{\partial v}{\partial t} \tag{2.2}$$

In this chapter, we will investigate the solution of these equations and look at some practical transmission lines.

The solution of Equations (2.1) and (2.2) proceeds along the line of differential equations in general, which the reader is assumed to have studied in other courses. This coupled system of two equations in terms of the voltage $v(z,t)$ and current $i(z,t)$ can be simplified by eliminating one variable from the two equations. For instance, by differentiating Equation (2.1) with respect to z and differentiating Equation (2.2) with respect to t, the equations can be combined to produce

$$\frac{\partial^2 v}{\partial z^2} = LC\frac{\partial^2 v}{\partial t^2} \tag{2.3}$$

Equation (2.3) is well-known to the math and science community and has been given the name "one-dimensional wave equation." In obtaining Equation (2.3), we eliminated the current from Equations (2.1) and (2.2); an alternate approach is to differentiate Equation (2.1) with respect to t and differentiate Equation (2.2) with respect to z to eliminate the voltage and obtain

$$\frac{\partial^2 i}{\partial z^2} = LC \frac{\partial^2 i}{\partial t^2} \tag{2.4}$$

Thus, the voltage and current satisfy the same second-order differential equation.

There are numerous ways to solve differential equations, and we fall back on a useful expedient: guess a solution and show that it works. The general solution we propose has the form

$$v(z,t) = V^+ f\left(t - \frac{z}{v_{\mathrm{p}}}\right) + V^- g\left(t + \frac{z}{v_{\mathrm{p}}}\right) \tag{2.5}$$

where V^+ and V^- are unknown coefficients (to be determined by initial or boundary conditions at the ends of the line) and f and g are arbitrary functions of one variable, also to be determined by initial or boundary conditions. By substituting Equation (2.5) into Equations (2.1) and (2.2), we determine that the current has the form

$$i(z,t) = \frac{V^+}{Z_0} f\left(t - \frac{z}{v_{\mathrm{p}}}\right) - \frac{V^-}{Z_0} g\left(t + \frac{z}{v_{\mathrm{p}}}\right) \tag{2.6}$$

In Equations (2.5) and (2.6), the parameter v_{p} has units of velocity (m/s) and is given by

$$v_{\mathrm{p}} = \frac{1}{\sqrt{LC}} \tag{2.7}$$

while Z_0 has units of resistance (Ω) and is given by

$$Z_0 = \sqrt{L/C} \tag{2.8}$$

The parameter Z_0 is known as the characteristic resistance, or more commonly, the characteristic impedance of the line.

To demonstrate that Equations (2.5) and (2.6) are the solutions to Equations (2.3) and (2.4), we simply substitute them into the equations. Consider just the part of the solution involving the coefficient V^+. By differentiating, we obtain

$$\frac{\partial^2 v}{\partial z^2} = \left(\frac{-1}{v_{\mathrm{p}}}\right)^2 V^+ f''\left(t - \frac{z}{v_{\mathrm{p}}}\right) \tag{2.9}$$

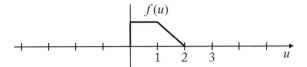

FIGURE 2.1: A function of limited support.

$$\frac{\partial^2 v}{\partial t^2} = V^+ f'' \left(t - \frac{z}{v_p} \right)$$

(2.10)

Clearly, this part of the solution satisfies Equation (2.3). The other part of the solution is easily checked, as is the current and equation (2.4).

Now that we have a solution, what does it mean? Or, to put it differently, what is a one-dimensional wave? In order to explore this question, we consider the following. Suppose we have a function $f(u)$, such as that given in Figure 2.1. We can easily construct $f(t + z/v_p)$ for specific values of space and time; Figure 2.2 shows three specific plots versus space (z) at fixed times for $v_p = 300$ m/s. From Figure 2.2, we observe that $f(t + z/v_p)$ represents a function that retains its shape as time marches forward but changes its position in z; specifically, it travels in the $-z$ direction with a constant velocity v_p. If we plot the function $f(t - z/v_p)$ as in Figure 2.3, we observe that it also retains its shape as it shifts along the z axis; it travels in the $+z$ direction with constant velocity v_p. In reading these figures, pay particular attention to the leading edge of the waveform and the trailing edge of the waveform. The leading edge is always the first part of the waveform to appear at a given location as time marches forward.

From the preceding example, we infer that a one-dimensional wave is a function that propagates with a constant velocity without changing its shape. Furthermore, the voltage and current

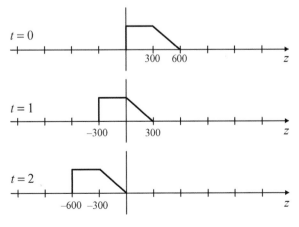

FIGURE 2.2: The function $f(t + z/300)$ for three values of t.

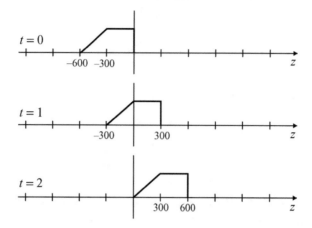

FIGURE 2.3: The function $f(t - z/300)$ for three values of t.

waveforms that satisfy the lossless transmission line equations are one-dimensional waves. The minus sign in the argument of the wave function indicates the direction: a plus sign means the wave is traveling in the $-z$ direction, while a minus sign indicates the wave is traveling in the $+z$ direction.

2.1   IDEAL PARALLEL STRIP TRANSMISSION LINE

We now turn our attention to several common types of transmission line. The simplest type of line might well be the ideal parallel strip transmission line illustrated in Figure 2.4. This line is "ideal" in the sense that the fringing nature of the fields at the open ends is ignored. The fringing fields are usually too severe to ignore, however, and shortly we will consider the microstrip line, which offers a better approximation to the true situation. Under lossless assumptions, the ideal strip line has an inductance per unit length of

$$L = \mu \frac{d}{w} \qquad (2.11)$$

and a capacitance per unit length of

$$C = \varepsilon \frac{w}{d} \qquad (2.12)$$

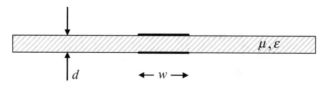

FIGURE 2.4: Ideal parallel strip transmission line.

where μ and ε are the total permeability and permittivity parameters, respectively, given by

$$\mu = (4\pi \times 10^{-7})\mu_r \qquad \text{(in H/m)} \tag{2.13}$$

$$\varepsilon = (8.854 \times 10^{-12})\varepsilon_r \qquad \text{(in F/m)} \tag{2.14}$$

Consequently, the propagation velocity of a signal on this line is given by

$$v_p = \frac{1}{\sqrt{LC}} = \frac{1}{\sqrt{\mu\varepsilon}} \tag{2.15}$$

while the characteristic impedance is given by

$$Z_0 = \sqrt{L/C} = \frac{d}{w}\sqrt{\mu/\varepsilon} \tag{2.16}$$

Observe that the propagation velocity of an electrical signal on this line is independent of the height and width of the line geometry. Furthermore, the velocity is the same as that in an infinite medium of the same material. This is always true if the line is constructed from a uniform (homogeneous) material, and is basically a physical constraint on the inductance per unit length and the capacitance per unit length that takes the form

$$LC = \mu\varepsilon \tag{2.17}$$

Thus, the dimensions of the metal strips play no role in the velocity of an electrical signal on the line. The characteristic impedance does depend on the geometry of the line (the cross-sectional dimensions) as well as the material. By adjusting the height and width of the strips, Z_0 can be set at any desired value. Note that Z_0 decreases as the width of the line, w, increases.

2.2 COAXIAL TRANSMISSION LINE

Figure 2.5 shows the cross section of a coaxial transmission line. Coaxial cables are widely used in both low- and high-frequency electrical systems since the outer conductor provides a shielding effect that reduces interference from nearby conductors.

Under lossless assumptions and neglecting the internal inductance of the conductors (usually small at high frequencies), the coaxial line has an inductance per unit length of

$$L = \frac{\mu}{2\pi}\ln(b/a) \tag{2.18}$$

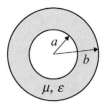

FIGURE 2.5: Coaxial cable cross section.

and a capacitance per unit length of

$$C = \frac{2\pi\varepsilon}{\ln(b/a)} \qquad (2.19)$$

where μ and ε are given by Equations (2.12) and (2.13). The propagation velocity of a signal on this line is

$$v_p = \frac{1}{\sqrt{LC}} = \frac{1}{\sqrt{\mu\varepsilon}} \qquad (2.20)$$

while the characteristic impedance is given by

$$Z_0 = \sqrt{L/C} = \frac{\ln(b/a)}{2\pi}\sqrt{\mu/\varepsilon} \qquad (2.21)$$

Again, the propagation velocity of an electrical signal on this line depends only on the material separating the conductors. Thus, the radii of the conductors play no role in the velocity. The characteristic impedance does depend on the cross-sectional dimensions, and can be set at a desired value by adjusting the ratio b/a.

We note in passing that microwave test equipment uses the standard coaxial characteristic impedance of 50 Ω, while the CATV industry has standardized on 75 Ω. The minimum loss for a given signal strength is realized near an impedance of 75 Ω, while the maximum power capability of an air-filled coaxial line is realized near 32 Ω. Since CATV distribution does not require high power levels, the 75-Ω standard is selected to minimize losses. The 50-Ω standard is a compromise between high power capability and low losses for an air-filled line and represents the maximum power impedance for a line constructed with a dielectric material having $\varepsilon_r = 2.5$.

2.3   MICROSTRIP TRANSMISSION LINE

The microstrip line shown in Figure 2.6 provides a useful approximation to a trace on a PWB. A signal consists of electric and magnetic fields between the conductors, which in actuality fringe around the conductor edges. The analysis of microstrip lines is made more difficult because of the

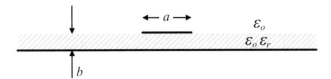

FIGURE 2.6: The cross section of a microstrip line.

fringing nature of the fields and the presence of two different materials around the conductors. The following expressions are inferred from measurement studies, and incorporate the fringing effect of the fields.

Since the fields around the two conductors extend throughout two different materials, it is useful to define an effective permittivity

$$\varepsilon_{\text{eff}} = 8.854 \times 10^{-12} \left[\frac{\varepsilon_r + 1}{2} + \frac{\varepsilon_r - 1}{2} \frac{1}{\sqrt{1 + 12b/a}} \right] \text{ (in F/m)} \tag{2.22}$$

which is a weighted average of the permittivity of the substrate and superstrate (air) regions. As a result, the velocity of an electrical signal on this line is given by

$$v_{\text{p}} = \frac{1}{\sqrt{\mu \varepsilon_{\text{eff}}}} \tag{2.23}$$

Relatively simple approximate expressions for the characteristic impedance are given by [1–2]

$$Z_0 = \frac{1}{2\pi} \sqrt{\mu/\varepsilon_{\text{eff}}} \ln \left(\frac{8b}{a} + \frac{a}{4b} \right), \quad a < b \tag{2.24}$$

$$Z_0 = \sqrt{\mu/\varepsilon_{\text{eff}}} \left(\frac{1}{a/b + 1.393 + 0.667\ln(a/b + 1.444)} \right), \quad a > b \tag{2.25}$$

These expressions can be inverted if it is desired to design a microstrip line with a particular impedance Z_0, using

$$\frac{a}{b} = \begin{cases} \dfrac{8e^A}{e^{2A} - 2}, & a < 2b \\ \dfrac{2}{\pi} \left\{ B - 1 - \ln(2B - 1) + \dfrac{\varepsilon_r - 1}{2\varepsilon_r} \left[\ln(B - 1) + 0.39 - \dfrac{0.61}{\varepsilon_r} \right] \right\}, & a > 2b \end{cases} \tag{2.26}$$

where ε_r is the relative permittivity of the substrate and

$$A = \frac{Z_0}{60}\sqrt{\frac{\varepsilon_r + 1}{2}} + \frac{\varepsilon_r - 1}{\varepsilon_r + 1}\left(0.23 + \frac{0.11}{\varepsilon_r}\right) \qquad (2.27)$$

$$B = \frac{377\pi}{2Z_0\sqrt{\varepsilon_r}} \qquad (2.28)$$

Additional information on microstrip lines and extensions of the above formulas for impedance under a wide range of conditions can be found in the literature [1].

Example: A microstrip configuration is designed with $a = 2$ mm, $b = 1$ mm, and $\varepsilon_r = 5$. Find v_p and Z_0 for this transmission line.

Solution: The effective permittivity is determined from Equation (2.22) as

$$\varepsilon_{\text{eff}} = \left(8.854 \times 10^{-12}\right)\left\{\frac{5 + 1}{2} + \frac{5 - 1}{2}\frac{1}{\sqrt{1 + 6}}\right\}$$

$$= 3.76\varepsilon_0$$

$$= 3.33 \times 10^{-11} \text{ F/m}$$

which results in a propagation velocity of

$$v_p = \frac{3 \times 10^8}{\sqrt{3.76}} = 1.55 \times 10^8 \text{ m/s}$$

Since $a > b$, the characteristic impedance can be determined from Equation (2.25):

$$Z_0 = \frac{376.7}{\sqrt{3.76}}\frac{1}{2 + 1.393 + 0.667\ln(2 + 1.444)} = 46 \, \Omega$$

2.4 SYMMETRICAL STRIPLINE TRANSMISSION LINE

For traces embedded within a PWB, module, or chip, a better model for the trace may be offered by the symmetrical stripline geometry shown in Figure 2.7. The stripline involves a trace between two ground planes, with a single medium between the ground planes, and thus there is no need to use a weighted average for ε_r. The expressions given below are based on measurements and include the fringing effect of the fields.

In common with other uniformly filled lines, the velocity of an electrical signal is given by

$$v_p = \frac{1}{\sqrt{\mu\varepsilon}} \qquad (2.29)$$

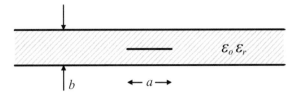

FIGURE 2.7: Cross section of a symmetrical stripline.

The characteristic impedance is given by

$$Z_0 = \frac{30\pi}{\sqrt{\varepsilon_r}} \frac{b}{a_{\text{eff}} + 0.441b} \tag{2.30}$$

where the parameter a_{eff} incorporates the fringing effect and is defined by

$$a_{\text{eff}} = \begin{cases} a, & a > 0.35b \\ a - \left(0.35 - \frac{a}{b}\right)^2 b, & a < 0.35b \end{cases} \tag{2.31}$$

As with the microstrip transmission line, the expression for characteristic impedance is approximate but is reported to be accurate to within about 1% for zero strip thickness and most common materials. More detailed consideration of stripline, and extensions to the nonsymmetrical case, may be found in the literature.

2.5   A NOTE ON FRINGING FIELDS

As a general rule, fringing fields are an undesired aspect of transmission line construction. Ideally, transmission lines carry most of their propagating signal power in *confined* fields — electric and magnetic fields internal to the dielectric substrate. The function of the line degrades when power is carried by *fringing* fields — fields that propagate in open air, near the surface of the transmission line material. Some transmission line geometries, such as the coaxial cable or the stripline configurations, completely confine the propagating fields to the interior of conductors. Other transmission line geometries may have fields that stray out into space, where they can act unpredictably with nearby objects.

Figure 2.8 depicts the field lines associated with several common transmission line geometries. Each transmission line consists of two conductors, and when these are excited with a voltage difference, the electric field lines emanate from the higher-voltage conductor to the lower-voltage conductor. The dielectric substrate provides insulation between the two conductors as well as mechanical stability.

While fringing fields are usually undesirable, field confinement comes with increased cost and an inconvenient form factor. Engineers would enjoy connecting everything with coaxial cables

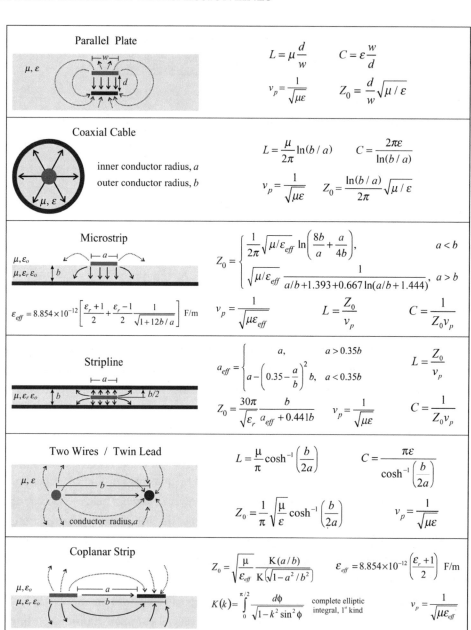

*Brian C. Wadell, *Transmission Line Design Handbook*, Artech House, Boston, MA, 1991.

FIGURE 2.8: Summary of common transmission lines and parameters.

in order to minimize crosstalk, loss, and interference. However, such an approach would not be practical for a computer processor, where hundreds of high-speed interconnections must coexist in a relatively small board space, nor would it be cost effective in terms of the required number of cables and connectors. While the stripline type of geometry offers improved field confinement, such an approach requires multilayer circuit boards that are more expensive to fabricate and more challenging to design than conventional single-layer microstrip boards.

Figure 2.8 also contains a summary of expressions for inductance and capacitance per unit length for the various transmission line geometries depicted and the characteristic impedance and propagation velocities for those lines.

In this chapter, several types of transmission lines and the mathematical form of the signals (one-dimensional waves) that can exist on those lines have been described. In subsequent chapters, these lines will be used to connect signal sources to receivers, under a range of circumstances. Chapter 3 considers the connection of DC batteries to resistive loads.

REFERENCES

[1]   C. A. Balanis, *Advanced Engineering Electromagnetics*. New York: Wiley, 1989.

[2]   D. M. Pozar, *Microwave Engineering*. New York: Wiley, 1998.

PROBLEMS

2.1   *Fill in the blanks:*

(a)   A _____ is a type of transmission line that is made by etching or milling away the top conductor on a conventional printed circuit board.

(b)   The _____ and the _____ are two examples of transmission lines that have complete field and signal confinement.

(c)   When the transit time of a printed circuit board causes delays in pulses that threaten the synchronization of a high speed circuit, we call this effect _____.

(d)   Three ways to increase Z_0 for a coaxial line are to _____ the radius of the inner conductor, _____ the radius of the outer conductor, or _____ the permittivity of the dielectric material.

2.2   *True or false:*

(a)   If two transmission lines have the same per-unit-length capacitance, they will have the same characteristic impedance Z_0.

(b)   The microstrip transmission line is made from a printed circuit board with etched traces above the dielectric and a ground plane on the bottom.

(c)   The characteristic impedance Z_0 represents the amount of power absorbed and converted to heat by the transmission line.

2.3 Using the function $f(u)$ from Figure 2.1, plot $f\left(t + \dfrac{z}{300}\right)$ and $f\left(t - \dfrac{z}{300}\right)$ versus time, at $z = -600$, 0, and +600 m.

2.4 Fill in the following table of transmission line parameters, where each row corresponds to a unique transmission line.

	L	C	Z_0 (Ω)	v_p (m/s)
a		1 μF/m	50	
b	μ_0	ε_0		
c			200	1×10^7
d	200 μH/m			4×10^7
e			75	2.4×10^8

2.5 The following represent solutions of the transmission line differential equations:

$$v(z, t) = 60 \cos(1000t - 5z) - 120\pi f\left(2t + \frac{z}{100} + 12\right) \text{ (in V)}$$

$$i(z, t) = 2\pi f\left(2t + \frac{z}{100} + 12\right) + \cos(1000t - 5z) \text{ (in A)}$$

for some arbitrary function $f(u)$. Answer the following questions about this solution:
(a) What is the forward-propagating voltage waveform?
(b) What is the backward-propagating current waveform?
(c) What is the characteristic impedance of this line?
(d) What is the velocity of propagation for a signal on this line?
(e) What is the total voltage at $t = 0$ at the input end of the line ($z = 0$)?

2.6 A bankrupt student discovers a 15-m length of coaxial cable in the laboratory with connectors on both ends, so it is not possible to see the center conductor. Another student bets her $50 that she cannot determine the outside diameter of the inner conductor or the nature of the insulating material between the conductors without cutting the cable open.

 The student takes the bet. First, she holds the cable up to a magnet and determines that since there is no attraction, the cable is constructed only of nonmagnetic materials with $\mu = \mu_0 = 4\pi \times 10^{-7}$ H/m. Second, she uses a dual trace oscilloscope to measure the time it takes a pulse to travel down the line and determines that it takes 70 ns. Third, she uses the capacitance meter on her digital voltmeter to determine the total capacitance of the 15-m cable to be 3 nF. Finally, by measuring the outer diameter of the cable and guessing at the thickness of the outer conductor, she estimates that the outer conductor has an inside radius of $b = 6$ mm. With this information, she is able to answer the following questions and win the bet! Now it is your turn:

(a) What is the propagation velocity on the transmission line?

(b) What is the relative permittivity of the insulating material between the conductors of the coax?

(c) What is the capacitance per unit length for the cable?

(d) What is the diameter of the center conductor?

(e) What is the characteristic impedance of the line?

2.7 Design a coaxial cable with characteristic impedance $Z_0 = 50\ \Omega$. The cable must have an outer conductor radius of 1 cm and a solid inner copper core of radius a. You must select a dielectric with relative permittivity ε_r and a value for the conductor core radius a that achieves the desired impedance while minimizing the cost-per-meter of the cable. Assume that copper conductor costs \$2000/m^3 and the dielectric material cost is given by the function \$$(200 + 25\varepsilon_r)$ / m^3. (Note that higher permittivity dielectrics are more expensive!) In addition to ε_r and a, determine the cost-per-foot of your optimum design as well as the velocity of propagation, the inductance per unit length, and the capacitance per unit length. *Hint:* It is probably easiest to plot the total cost versus ε_r and find the optimal design parameters by a visual inspection of the graph.

2.8 Design a microstrip transmission line with $Z_0 = 100\ \Omega$ that will be etched onto a dielectric substrate with $\varepsilon_r = 3$ and a thickness of 4 mm. What should be the width of the microstrip line?

2.9 The following represent invalid solutions of the transmission line differential equations. Explain why each is impossible for a linear transmission line in terms of basic physical properties (i.e., characteristic impedance, pulse shape, attenuation, velocity of propagation, etc.). Do not give purely mathematical answers!

(a) $v(z, t) = 100 \cos\left(2\pi ft - z\right) + 50 \sin\left(2\pi ft + z\right)$

$i(z, t) = 5 \cos\left(2\pi ft - z\right) - 5 \sin\left(2\pi ft + z\right)$

(b) $v(z, t) = 20 \dfrac{\sin\left(t - 5z\right)}{t - 5z}$

$i(z, t) = \sin\left(t - 5z\right)$

(c) $v(z, t) = 75 \exp\left(-|t - z|^2\right) + 75 \exp\left(-\left|t + \dfrac{z}{2}\right|^2\right)$

$i(z, t) = \exp\left(-|t - z|^2\right) - \exp\left(-\left|t + \dfrac{z}{2}\right|^2\right)$

CHAPTER 3

DC Signals on a Resistively Loaded Transmission Line

Objectives: Walk through the evolution of a transmission line problem for a line terminated with resistive loads. Introduce concepts of reflection coefficient, equivalent circuits for the line, and the reflection or bounce diagram.

Consider a transmission line with circuits attached to each end, as shown in Figure 3.1. We have constructed a system containing a DC battery, a resistor representing the internal resistance of the battery, and a load resistor at the end. The battery end of the line is usually referred to as the "input" or "generator" end, while the other end is known as the "load" end. Specific values are given for the resistors and the battery voltage so that we can work things numerically. The transmission line is depicted as two parallel wires for convenience; we will draw it this way even though it might represent a coaxial cable or some other line geometry. The transmission line in Figure 3.1 has characteristic impedance of 75 Ω, and rather than specify the propagation velocity, we will instead work with a time delay T, representing the one-way transit time of the line (if the velocity is v_p and the line length is L, then $T = L/v_p$).

The switch in Figure 3.1 is closed at time $t = 0$. Immediately after the switch is closed, a voltage appears across the input end of the line. This voltage causes a wave to start to travel down

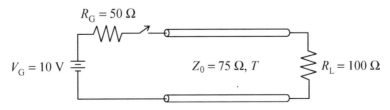

FIGURE 3.1: A transmission line system.

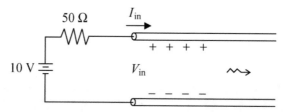

FIGURE 3.2: The initial wave launched on the line.

the transmission line, as depicted in Figure 3.2. Since the battery and resistor comprise a traditional circuit, Kirchhoff's laws describe the voltage and current in the circuit at the input end. The voltage and current waves on the line itself, however, are described by the transmission line equations and solution discussed in Chapters 1 and 2. There is no signal impinging from the load end of the line, and consequently, the line voltage and current at the input end at time $t = 0^+$ are given by a forward-going wave with

$$v^+(z, t) = V^+ f\left(t - \frac{z}{v_p}\right) \tag{3.1}$$

$$i^+(z, t) = \frac{V^+}{Z_0} f\left(t - \frac{z}{v_p}\right) \tag{3.2}$$

Since the excitation in this example is DC, the function f in these equations is just a unit step function $u(t)$ and will be omitted from the following equations in favor of the coefficient V^+. At $t = 0^+$, the voltage and current across the input end of the line are related by

$$V_{in} = V^+ \tag{3.3}$$

$$I_{in} = V^+/Z_0 \tag{3.4}$$

Thus, $I_{in} = V_{in}/Z_0$, and the line looks like a resistor with resistance Z_0.[1] An equivalent circuit for the input end of the line at time $t = 0^+$ can be constructed as shown in Figure 3.3.

From the equivalent circuit, we determine that

$$V_{in} = V_G \frac{Z_0}{Z_0 + R_G} = 10 \frac{75}{75 + 50} = 6.0 \text{ V} \tag{3.5}$$

---

[1]The line only appears to be a resistive load; it accepts power that propagates down the line. It is important to note that the line is lossless and does not dissipate energy like an actual resistor.

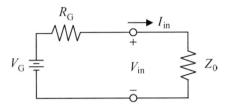

FIGURE 3.3: Equivalent circuit for generator end of the line at $t = 0^+$.

$$I_{in} = V_{in}/Z_0 = 6/75 = 0.08 \text{ A} \tag{3.6}$$

Once the wave has been launched from the input end of the line, its leading edge propagates down the line with velocity v_p, until it reaches the load end of the line at time $t = T$.

Figure 3.4 displays snapshots of the instantaneous voltage as a function of position for several values of t. There is also a current waveform with the same shape, with the ratio of voltage to current fixed at 75 Ω in accordance with Equations (3.1) and (3.2).

At time $t = T$, the leading edge of the waveform reaches the load end of the line, shown in Figure 3.5, and produces a voltage and current across the load resistor R_L. Since $R_L = 100 \ \Omega \neq Z_0$, Ohm's law is not satisfied by R_L, $v^+(z,t)$ and $i^+(z,t)$. To satisfy Ohm's law at the load, as well as the transmission line equations, a reflected wave is generated at the load end of the line, according to the general solution of the transmission line equations, with

$$v^-(z, t) = V^- g\left(t + \frac{z}{v_p}\right) \tag{3.7}$$

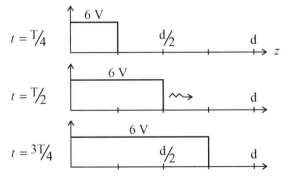

FIGURE 3.4: Snapshots of the instantaneous voltage along the line.

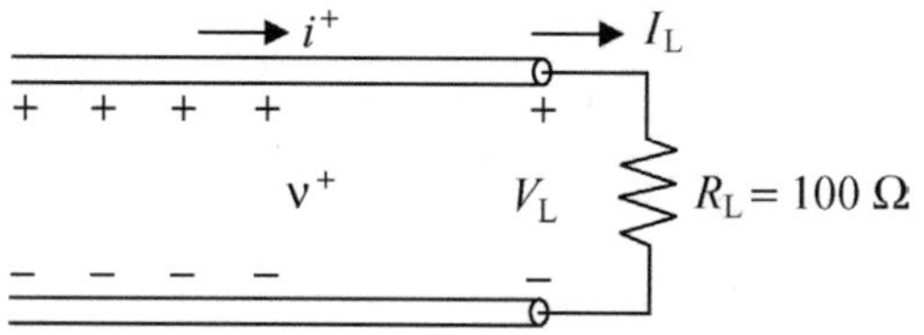

FIGURE 3.5: The leading edge of the waveform approaching the load end of the line.

$$i^-(z,t) = -\frac{V^-}{Z_0}g\left(t + \frac{z}{v_p}\right) \tag{3.8}$$

Again, the function g is a unit step function $u(t - T)$ and will be ignored in the equations to follow. At time $t = T^+$, the load voltage and current must satisfy the transmission line equations

$$V_L = V^+ + V^- \tag{3.9}$$

$$I_L = \frac{V^+}{Z_0} - \frac{V^-}{Z_0} \tag{3.10}$$

as well as Ohm's law

$$I_L = \frac{V_L}{R_L} \tag{3.11}$$

Combining Equations (3.9) and (3.11) yields

$$I_L = \frac{V^+}{R_L} + \frac{V^-}{R_L} \tag{3.12}$$

Equations (3.10) and (3.12) can be solved simultaneously to produce

$$V^- = \Gamma_L V^+ \tag{3.13}$$

where Γ_L is the reflection coefficient at the load, defined by

$$\Gamma_L = \frac{R_L - Z_0}{R_L + Z_0} = \frac{100 - 75}{100 + 75} = \frac{1}{7} \tag{3.14}$$

The reflection coefficient is a ratio of two voltages and therefore is unitless. From Equation (3.13), we determine that $V^- = 0.857$ V and $I^- = -V^-/Z_0 = -0.0114$ A.

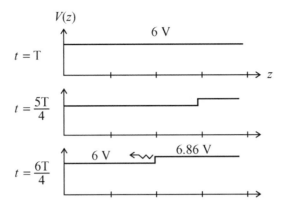

FIGURE 3.6: Snapshots of the line voltage for $T < t < 1.5\ T$.

The reflected wave propagates in the $-z$ direction with velocity v_p, until it reaches the input end of the line at time $t = 2T$. Figure 3.6 shows a series of snapshots of the voltage as a function of position between time $t = T$ and $t = 2T$. During this time interval, the line voltage gradually changes from 6.0 to 6.857 V. However, at $t = T$, the load voltage immediately changes from $V_\mathrm{L} = 0$ to $V_\mathrm{L} = V^+ + V^- = 6.857$ V. It is important to note that the load does not "see" the intermediate value of 6.0 V!

During the time interval $0 < t < T$, there is only a single waveform propagating in the $+z$ direction on the line. From $t = T$ to $t = 2T$, there are waves traveling in both directions on the line. Waves traveling in the $+z$ direction have the ratio of voltage to current fixed at Z_0, while waves traveling in the $-z$ direction have their ratio fixed at $-Z_0$. (The significance of the minus sign is that the reference direction of the current is always taken to be the $+z$ direction, while in fact the reflected current flows in the opposite direction.) During the time interval from $t = T$ to $t = 2T$, the load voltage and current are constant, with values given in this case by

$$V_\mathrm{L} = V^+ + V^- = 6.857 \text{ V} \qquad (3.15)$$

$$I_\mathrm{L} = I^+ + I^- = 0.0686 \text{ A} \qquad (3.16)$$

From the equations for the load end of the line, Equations (3.9) to (3.12), we also determine that

$$V_\mathrm{L} = V^+ \frac{2R_L}{R_L + Z_0} \qquad (3.17)$$

or, equivalently,

$$V_\mathrm{L} = V^+ (1 + \Gamma_\mathrm{L}) \qquad (3.18)$$

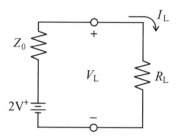

FIGURE 3.7: Equivalent circuit for the load end of the line at time $t = T^+$.

From Equation (3.17), it is straightforward to back out the Thevenin equivalent circuit for the load end of the transmission line at time $t = T^+$ (Figure 3.7). Note the somewhat counterintuitive factor of 2 for the equivalent voltage!

At time $t = 2T$, the leading edge of the reflected waveform arrives at the generator end of the line, where it encounters another impedance mismatch since $R_G \neq Z_0$. Using an analysis similar to that employed above, we introduce a reflection coefficient at the generator

$$\Gamma_G = \frac{R_G - Z_0}{R_G + Z_0} = \frac{50 - 75}{50 + 75} = -\frac{1}{5} \tag{3.19}$$

associated with a new incremental positive-going waveform having voltage

$$V^{++} = \Gamma_G V^- = (-0.2)(0.857) = -0.171 \text{ V} \tag{3.20}$$

Figure 3.8 illustrates the propagation of this transient along the transmission line from $t = 2T$ to $t = 3T$. During this interval, we can think of three separate waveforms superimposed on the line, the incident wave, the reflected wave, and the re-reflected signal.

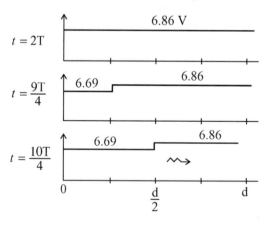

FIGURE 3.8: Snapshots of the line voltage for $2T < t < 3T$.

We observe that the reflection coefficients arising in this example are less than 1.0, and therefore, the reflections tend to die out over time. In fact, one can show that

$$-1 \leq \Gamma = \frac{R - Z_0}{R + Z_0} \leq 1 \tag{3.21}$$

for any nonnegative values of R and Z_0. If the reflection coefficient magnitude is less than 1.0, the line voltage and current will stabilize at a steady DC value given by the infinite summation

$$V = V^+ \{1 + \Gamma_L + \Gamma_L \Gamma_G + \Gamma_L \Gamma_G \Gamma_L + \cdots\}$$
$$= V^+ (1 + \Gamma_L) \sum_{n=0}^{\infty} (\Gamma_L \Gamma_G)^n \tag{3.22}$$
$$= V^+ (1 + \Gamma_L) \frac{1}{1 - \Gamma_L \Gamma_G}$$

where the geometric series converges whenever the magnitude of the product $\Gamma_L \Gamma_G$ is less than one. Therefore, the steady-state DC voltage along the line as $t \to \infty$ is

$$V_{in} = V_L = V = V_G \left(\frac{Z_0}{Z_0 + R_G} \right) (1 + \Gamma_L) \frac{1}{1 - \Gamma_L \Gamma_G} \tag{3.23}$$
$$= V_G \left(\frac{R_L}{R_L + R_G} \right)$$

An equivalent circuit for the line as $t \to \infty$ is shown in Figure 3.9. (Since the transmission line is energized to a single voltage and current as $t \to \infty$, it may be completely removed from the original circuit without changing the other voltages or currents.) For this example, $V \to 6.667$ V and $I \to 0.0667$ A as $t \to \infty$.

In order to keep track of the various quantities arising in the transmission line analysis, it is convenient to plot the voltage or current on a two-dimensional axis showing position on the line versus time. Such a plot is known as a *reflection diagram* or bounce diagram. The key elements of

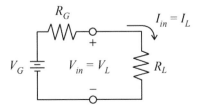

FIGURE 3.9: An equivalent circuit for the line as $t \to \infty$.

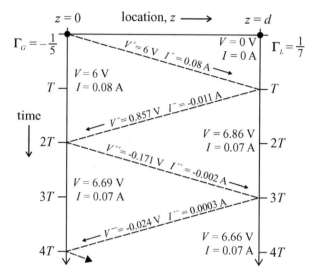

FIGURE 3.10: A reflection diagram.

the plot are the leading edge of the waveforms propagating along the line, which appear as straight lines. Figure 3.10 shows a portion of the reflection diagram for the previous example.

The reflection diagram makes the construction of the solution very systematic. Only three values are needed: Γ_L, Γ_G, and the initial incident voltage. The incremental voltages are obtained from the product of the previous incremental voltage and the reflection coefficients. On the diagram, the total voltages and total currents may be displayed in the triangles where they remain constant. These are the sum of all the previous incremental values. Along the straight lines representing the leading edge of the waveform, it is convenient to record the incremental voltages and currents.

If it is desired to graph the voltage or current at some time as a function of position on the line, a horizontal cut through the reflection diagram provides the plot. A plot of the voltage or current versus time at some location is a vertical cut through the diagram. For example, Figure 3.11 shows a plot of the voltage at a location three-quarters of the way down the line. As another example, a plot of the load voltage can be obtained from the extreme right side of the diagram.

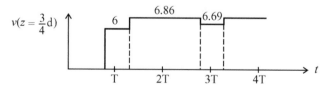

FIGURE 3.11: A plot of the voltage at $z = 0.75d$, as obtained from the reflection diagram.

In summary, we have walked through an example involving a DC generator driving a transmission line with resistive loads. Transmission line theory allows us to analyze any uniform transmission line in a similar manner, given the characteristic impedance Z_0 and the time delay T of the line. By arranging the results on a reflection diagram, we can systematically track the incremental changes until reaching a point where the system voltages and currents stabilize.

PROBLEMS

3.1   *Fill in the blank:* Complete reflection occurs if the load is either a _____ circuit or an _____ circuit.

3.2   *Fill in the blank:* The reflection coefficients of an open circuit, short circuit, and matched circuit are _____, _____, and _____, respectively.

3.3   Mona is a graduate teaching assistant in an electronics laboratory. One day she notices two large canisters in the corner of the lab: one contains thousands of 100-nF capacitors and the other contains thousands of 1-mH inductors. Out of sheer boredom, she connects all of the capacitors and inductors together to form a long, repetitive chain:

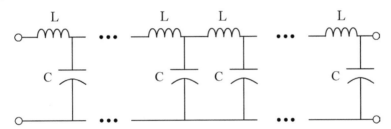

The chain is so long that it wraps around the hallway for several hundred feet. Mona then connects one end to an ideal (zero-impedance) function generator and excites the chain with a single 100-V square pulse. She hooks up numerous voltmeters, ammeters, and oscilloscopes along the chain to observe what happens.

Describe the behavior that Mona observes, including numerical values for voltages and currents measured in the chain.

3.4   The transmission line in the system below has a one-way transit delay of T seconds. The line is completely discharged for $t \leq 0$.

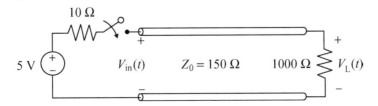

The switch is closed at $t = 0$. Plot the load voltage $V_L(t)$ as a function of time during the interval $0 < t < 10T$, clearly labeling all voltage levels.

3.5 Consider an initially uncharged transmission line with transit time T, generator reflection coefficient $\Gamma_G = -\frac{1}{2}$, and load reflection coefficient $\Gamma_L = \frac{1}{2}$. At $t = 0$, a DC source of 16 V is connected to the generator end of the line. Plot the voltage observed at the input end of the line, at the load end of the line, and at a point three-quarters of the distance to the load, for the time interval $0 < t < 6T$.

3.6 In the system shown below, the line has a one-way transit delay of T seconds. The line is completely discharged for $t \leq 0$.

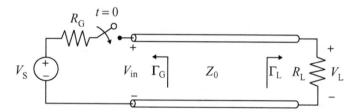

For the following source and component values, construct a reflection diagram for the system and plot the input voltage V_{in} and the load voltage V_L for the time interval $0 < t < 6T$:

(a) $V_S = 12$ V, $R_G = 25 \ \Omega$, $Z_0 = 50 \ \Omega$, and $R_L = 200 \ \Omega$.

(b) $V_S = 12$ V, $R_G = 25 \ \Omega$, $Z_0 = 50 \ \Omega$, and $R_L = 20 \ \Omega$.

(c) $V_S = 12$ V, $R_G = 25 \ \Omega$, $Z_0 = 50 \ \Omega$, and $R_L = 50 \ \Omega$.

(d) $V_S = 12$ V, $R_G = 50 \ \Omega$, $Z_0 = 50 \ \Omega$, and $R_L = 100 \ \Omega$.

(e) $V_S = 12$ V, $R_G = 50 \ \Omega$, $Z_0 = 50 \ \Omega$, and $R_L = 20 \ \Omega$.

3.7 Repeat Problem 3.6, but for each case, plot instead the input current I_{in} and load current I_L for the time interval $0 < t < 6T$.

CHAPTER 4

Termination Schemes

Objectives: Continue to illustrate the application of transmission line theory to resistively loaded lines excited by DC generators, using examples. Introduce power definitions, and present a variety of termination schemes that eliminate or reduce reflections from mismatched loads.

Consider the transmission line system shown in Figure 4.1, containing a DC battery, a resistor representing the internal resistance of the battery, and a short-circuit load at the end. The transmission line has characteristic impedance of 50 Ω, and one-way time delay T. It is a straightforward process to calculate the reflection coefficients

$$\Gamma_L = \frac{R_L - Z_0}{R_L + Z_0} = \frac{0 - 50}{0 + 50} = -1 \tag{4.1}$$

$$\Gamma_G = \frac{R_G - Z_0}{R_G + Z_0} = \frac{100 - 50}{100 + 50} = \frac{1}{3} \tag{4.2}$$

and the initial voltage excited on the line,

$$V_{in} = V_G \frac{Z_0}{Z_0 + R_G} = 300 \frac{50}{50 + 100} = 100 \text{ V} \tag{4.3}$$

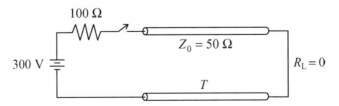

FIGURE 4.1: A transmission line system. The switch is closed at $t = 0$.

Using these three parameters, we can construct the reflection diagram. Figure 4.2 shows the reflection diagram constructed for the system of Figure 4.1.

There are several noteworthy aspects of this example. First, observe that the voltages appearing in the triangles along the load end of the line are always zero. The zero voltage is a consequence of the short circuit load, and gives us a means to quickly check the correctness of our work. Second, we note that although the voltage across the load is zero, the current through the load is not. As $t \to \infty$,

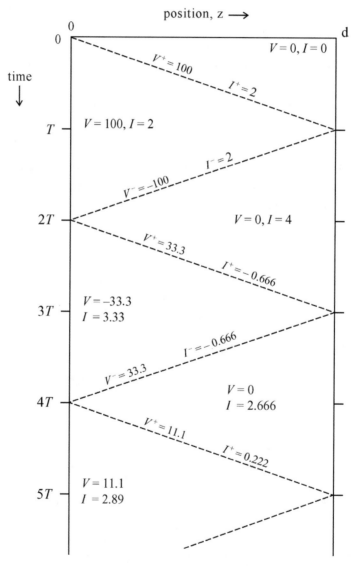

FIGURE 4.2: The reflection diagram for the system in Figure 4.1.

the line voltage stabilizes at 0 V and the current stabilizes at 3.0 A. Observe that even after just a couple of reflections, the voltage and current are approaching their steady-state values.

The power carried by the initial waveform traveling in the +z direction is given by

$$P^+ = V^+I^+ = (100)(2) = 200 \text{ W} \qquad (4.4)$$

The power carried by the first reflected wave is

$$P^- = V^-I^- = (-100)(2) = -200 \text{ W} \qquad (4.5)$$

Since the line is lossless and terminated by a short circuit, there is no power-absorbing mechanism, and all the incident power is reflected back toward the generator resistor. Alternative expressions for the powers are

$$P^+ = \frac{(V^+)^2}{Z_0} \qquad (4.6)$$

$$P^- = -\frac{(V^-)^2}{Z_0} \qquad (4.7)$$

where the minus sign indicates that the reference direction for power flow is actually in the +z direction.

Figure 4.3 shows a second transmission line system, this time with an open-circuited load. Part of the associated reflection diagram is given in Figure 4.4 (fill in the missing values yourself). In this situation, the load current is always zero, while the load voltage is nonzero. What are the values of voltage and current as $t \to \infty$?

We have studied several examples involving reflections occurring on mismatched transmission lines. It is important to realize that a similar behavior is exhibited by every electrical circuit, not just those containing coaxial cables or other uniform transmission lines. As an extreme example, consider the act of turning on an incandescent light by throwing a switch on the wall of a room. After the switch is closed, charges (electrons) move in the wires. The resulting current is perturbed

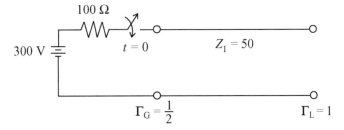

FIGURE 4.3: A transmission line system with an open-circuit load.

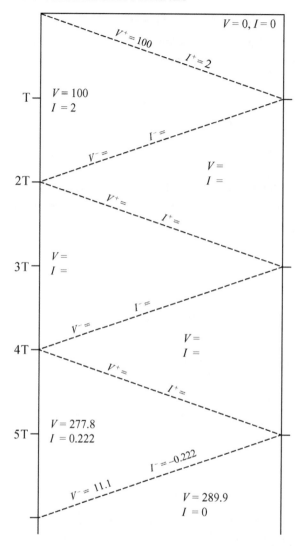

FIGURE 4.4: The partially filled reflection diagram for the system in Figure 4.3.

by any impedance mismatches encountered in the path between the switch and the light, and a mismatch at the light itself will also cause reflections. Of course, given the typical dimensions of a room (10 ft) and the speed of light in air (1 ft/ns), even a severe impedance mismatch will excite reflections that exist only during the first 50–100 nanoseconds. The human eye would not catch the flickering of the light for that brief an interval, even if the bulb did not require a far more appreciable time to heat up to its operating temperature and begin to illuminate. Thus, as a practical matter for that application, the transient reflections can be completely neglected and the $t \rightarrow \infty$ steady-state

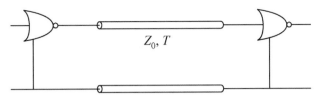

FIGURE 4.5: Two logic gates connected by a transmission line.

situation used instead. This is the same as using ordinary circuit theory with the transmission line treated as electrically short wires.

In other situations, however, the effects are more severe. Consider Figure 4.5, which shows two logic gates connected by a trace of some sort that we will model as a transmission line with Z_0 = 50 Ω and time delay T. We will also simplify the problem by assuming that when used as a driver, the logic gates can be replaced by an equivalent circuit consisting of a 5-V source in series with a 5-Ω resistor; when used at the receive end of the line, they can be replaced by an equivalent input resistance of 2000 Ω.[1] (In subsequent chapters, we will consider more sophisticated models of logic gates, incorporating capacitance as well as nonlinear effects.)

Figure 4.6 shows the equivalent problem, now in a form that should look familiar to the readers. It is readily determined that

$$\Gamma_L = \frac{R_L - Z_0}{R_L + Z_0} = 0.951 \tag{4.8}$$

$$\Gamma_G = \frac{R_G - Z_0}{R_G + Z_0} = -0.818 \tag{4.9}$$

$$V_{in} = V_G \frac{Z_0}{Z_0 + R_G} = 4.545 \text{ V} \tag{4.10}$$

Assuming that the initial state is zero voltage on the line, Figure 4.7 shows a portion of the reflection diagram constructed for the case where the driver turns on. The severe mismatches at both ends of the line combine to yield a slowly converging steady-state result in this situation. In fact, it takes until $t = 9T$ for the load voltage to exceed 3.5 V, which might be the minimum value necessary to reliably trigger the gate, without fluctuating back below that voltage level. Eventually, as $t \to \infty$, the load voltage stabilizes at 4.998 V.

---

[1]This example, and the ensuing discussion of termination schemes, has been repeated from Ref. [1].

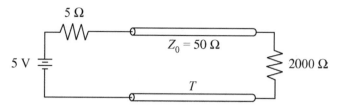

FIGURE 4.6: The equivalent transmission line system of Figure 4.5.

Clearly, the severe mismatches in the previous example can prevent a high-performance digital circuit from operating properly. How can these mismatches be eliminated or reduced so that they do not detrimentally affect the performance of a system? The process of eliminating these reflections is known as *matching* (or *impedance matching*) the line, and there are a variety of ways in which it can be accomplished.

In traditional high-frequency transmission line applications, the common approach is to match the line at both ends by adjusting the generator and load resistors in order that they each equal Z_0. For the example of Figure 4.6, a series resistor of 45 Ω could be added at the generator end, while a shunt resistor of 51.3 Ω could be added at the load end. The effect of these additional resistors is to make both Γ_G and Γ_L vanish, eliminating reflections. Figure 4.8 shows the resulting system and the reflection diagram. Unfortunately, we immediately observe a problem evident in Figure 4.8 — the voltage across the load is only 2.5 V, half of the generator voltage! Thus, this approach would not always work for digital applications, where the supply voltage might be limited to 5 V or less. (In traditional high-frequency applications, the generator voltages are adjusted to compensate for this factor of two.)

Alternate matching schemes are possible and might be more practical for the digital application. One approach is the *series termination*, illustrated in Figure 4.9. In the series termination scheme, a series matching resistor is inserted at the generator end of the line, while the load end is left as is. Consequently, for the example of Figure 4.6, the reflection coefficient at the generator is changed to

$$\Gamma_G = 0 \tag{4.11}$$

while the reflection at the load remains unchanged,

$$\Gamma_L = 0.951 \tag{4.12}$$

The series matching resistor changes the initial voltage to

$$V_{\text{in}} = 2.5 \text{ V} \tag{4.13}$$

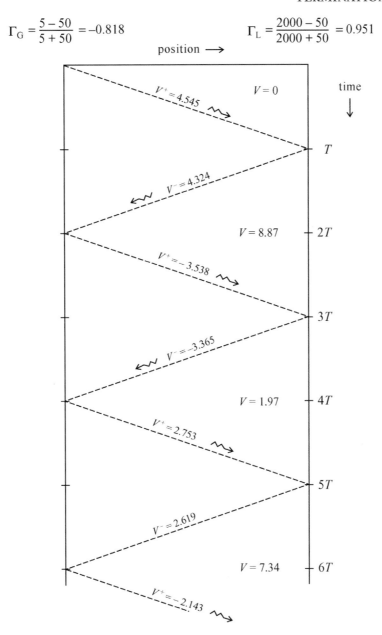

$$\Gamma_G = \frac{5 - 50}{5 + 50} = -0.818 \qquad\qquad \Gamma_L = \frac{2000 - 50}{2000 + 50} = 0.951$$

position ⟶

$V^+ = 4.545$ $V = 0$ time

T

$V^- = 4.324$

$V = 8.87$ $2T$

$V^+ = -3.538$

$3T$

$V^- = -3.365$

$V = 1.97$ $4T$

$V^+ = 2.753$

$5T$

$V^- = 2.619$

$V = 7.34$ $6T$

$V^+ = -2.143$

FIGURE 4.7: The reflection diagram for the system in Figure 4.6.

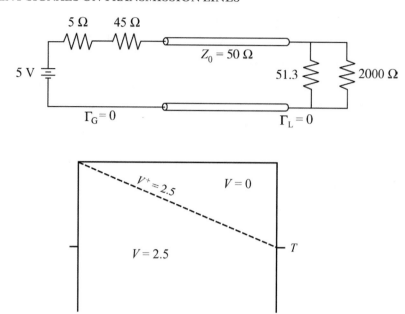

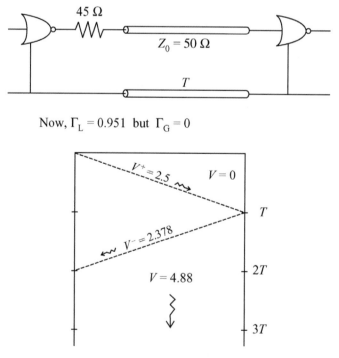

FIGURE 4.8: The system of Figure 4.6 with matching resistors added at both ends and the resulting reflection diagram.

Now, $\Gamma_L = 0.951$ but $\Gamma_G = 0$

FIGURE 4.9: A series termination applied to the system of Figures 4.5 and 4.6.

The reflection diagram is shown in Figure 4.9 and illustrates the result. Despite the severe reflection at the load end, the voltage at the load jumps from 0 to 4.88 V at $t = T$ and stays at that value. In fact, the reflection at the load is necessary in this case to provide voltage doubling, since the incident voltage is now only 2.5 V.

The second approach to matching is the *parallel termination*, illustrated in Figure 4.10. In the parallel termination scheme, a shunt matching resistor is inserted at the load end of the line, while the generator end is left mismatched. Consequently, for the example of Figure 4.6, the reflection coefficient at the generator remains

$$\Gamma_G = -0.818 \tag{4.14}$$

while the reflection at the load is eliminated,

$$\Gamma_L = 0 \tag{4.15}$$

The initial line voltage remains

$$V_{in} = 4.545 \text{ V} \tag{4.16}$$

The reflection diagram is shown in Figure 4.10 and illustrates the result. The mismatch at the generator end never enters into the analysis, since there is no negative-going wave on the line. (We

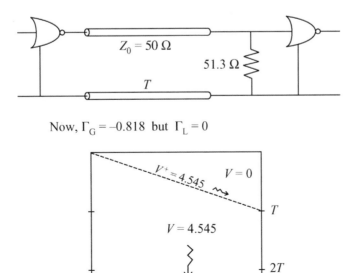

FIGURE 4.10: A parallel termination applied to the system of Figures 4.5 and 4.6.

defer until Chapter 6 the case where the driver changes from a high state to a low state, but in that situation, the parallel matching resistor also eliminates the problem with reflections.)

An important aspect of the series versus parallel matching approach is the power dissipated by the matching resistor. In the series matching case, once the line is charged, only a small amount of current flows through the 45-Ω matching resistor

$$I_{in} = (5.0 - 4.88)/50 = 2.4 \text{ mA} \tag{4.17}$$

meaning that the matching resistor dissipates 0.26 mW of power. In the parallel matching case, the total load current is

$$I_{L} = 4.545/50 = 90.9 \text{ mA} \tag{4.18}$$

meaning that the 51.3-Ω matching resistor dissipates 403 mW of power. We note that the lower power demand of the series match is more suitable for CMOS logic, the most common digital circuitry in use today. It is also worth noting that the parallel matching approach might be the preferred approach when used with high-performance emitter-coupled logic because it can play a dual role as a matching resistor for the transmission line and a pull-down resistor for the device circuitry.

If the power dissipation associated with a parallel matching resistor is a problem, an alternate approach employs an RC termination (Figure 4.11). If the time constant is large compared with T, the capacitor will charge slowly, cutting off the steady-state power dissipation without significantly affecting the load voltage. (The treatment of reactive loads will be discussed in more detail in Chapter 8.)

If the transmission line is terminated with an active device, it might be convenient to modify the device circuitry to provide matching instead of adding resistors (in other words, fabricate a matching network on the chip itself). An example of this, as well as a third approach to matching,

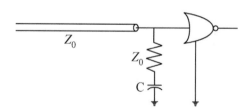

FIGURE 4.11: An RC termination scheme.

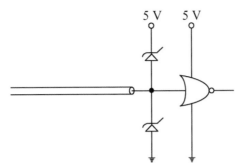

FIGURE 4.12: A diode termination scheme.

is the *diode termination* shown in Figure 4.12. The pair of diodes illustrated at the load end of the line hold the voltage within a set range and limit the overshoot and undershoot possible with a mismatched line.

As a specific example, suppose Zener diodes are employed with the ideal characteristic shown in Figure 4.13. The equivalent transmission line system and reflection diagram are shown in Figure 4.14. The presence of the diodes limits the overshoot to an extent that the voltage across the load stabilizes at time $t = T$ to within 0.8 V of the desired value.

In this lecture, we have considered several different termination schemes in an attempt to reduce or eliminate reflections caused by impedance mismatches. In the following lecture, we study the application of some of these ideas to transmission line configurations involving cascaded lines and fan-outs.

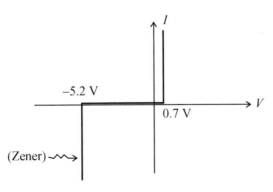

FIGURE 4.13: A possible Zener diode characteristic.

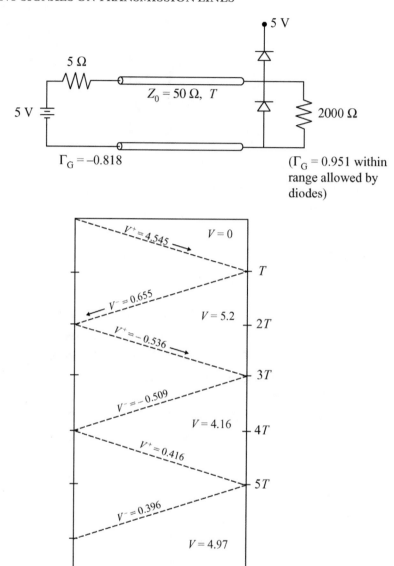

FIGURE 4.14: Application of the diode termination scheme to the system of Figure 4.6.

REFERENCE

[1]   M. Swaminathan, A. F. Peterson, and J. Kim, "Fundamentals of Electrical Package Design," in *Fundamentals of Microsystems Packaging*, R. R. Tummala, Ed. New York: McGraw-Hill, pp. 120–183, 2001.

PROBLEMS

4.1   *Fill in the blank*: If a load or source resistance is equal to the characteristic impedance of the transmission line, we say that the line is _____.

4.2   List 4 ways to terminate a transmission line to minimize reflections.

4.3   A transmission line system has a DC source voltage $V_S = 10$ V, $R_G = 20$ Ω, $Z_0 = 100$ Ω, and $R_L = 1800$ Ω. If the uncharged line is suddenly connected to the source, what is the initial power expended by the DC source? What is the power expended by the DC source upon reaching the steady state when $t \to \infty$? Where did the extra power go?

4.4   In the system shown below, the line has a one-way transit delay of T seconds. The line is completely discharged for $t \le 0$.

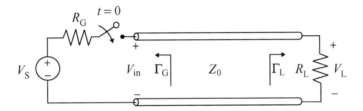

For $V_S = 12$ V, $R_G = 25$ Ω, $Z_0 = 50$ Ω, and $R_L = 100$ Ω, construct a reflection diagram for the system for the time interval $0 < t < 6T$. Use this diagram to plot the input power and the power dissipated in the load during this time interval.

4.5   The system shown below involves a peculiar circuit containing an ideal operational amplifier at the load of a transmission line:

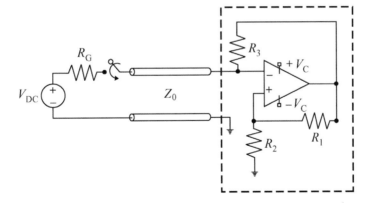

Because an operational amplifier is an active device, it is capable of supplying more power than it absorbs from the signaling network. For example, the Thevenin equivalent circuit for the load above is actually a pure resistance with a negative value:

$$R_L = -\left(\frac{R_2 R_3}{R_1 + R_2}\right)$$

Using basic principles of transmission line theory, explain qualitatively how this circuit will behave when the switch is thrown.

4.6 The dashed box in the transmission line system below contains an unusual matching circuit for switched DC logic signals. Explain why the circuit matches the load, as well as any limitations on the values of R_L that can be used in this type of circuit.

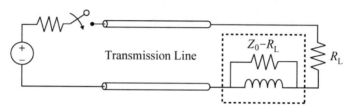

4.7 Consider a logic circuit consisting of two inverters connected through a transmission line. The equivalent circuit for the driver inverter consists of a 2.5-V DC source, a series resistor with $R_G = 10\ \Omega$, and a switch. The receiver inverter acts as a load of $R_L = 1\ k\Omega$. The transmission line has $Z_0 = 50\ \Omega$, and a one-way transit time delay of $T = 1$ ns. The receiver inverter gate is triggered when the load voltage reaches 2 V. Your goal is to redesign this system to eliminate reflections and minimize power dissipation.

(a) Develop a reflection diagram for the initial system described above.

(b) Add appropriate resistors to the input end and load end of the transmission line in order to eliminate reflections at both ends. What are the values for these resistors? How much power is dissipated when the line is fully charged? What is the major problem with the resulting design?

(c) Use an appropriate matching resistor only at the load end of the transmission line to eliminate reflections at the load. If the switch at the input end of the line is closed at $t = 0$, sketch the load voltage for $0 < t < 3T$. How much power is dissipated when the line is fully charged?

(d) Use an appropriate matching resistor only at the input end of the transmission line to eliminate reflections at the input. If the switch at the input end of the line is closed at $t = 0$, sketch the load voltage for $0 < t < 3T$. How much power is dissipated when the line is fully charged?

(e) Which design in (b), (c), and (d) is the best? Why?

CHAPTER 5

Equivalent Circuits, Cascaded Lines, and Fan-Outs

Objectives: Introduce Thevenin equivalent circuits for transmission lines. Extend the application of transmission line theory to cascaded lines excited with DC sources. Introduce transmission coefficients. Consider fan-outs and the application of matching techniques to these situations.

5.1   THEVENIN EQUIVALENT CIRCUITS

Consider the load end of a transmission line of characteristic impedance Z_0, terminated in some arbitrary resistive load, and excited at the generator end with a battery that produces DC waves on the line. During some time interval, the forward and reverse waves on the line satisfy the equations

$$V^+ + V^- = V_L \tag{5.1}$$

$$\frac{V^+}{Z_0} - \frac{V^-}{Z_0} = I_L \tag{5.2}$$

where V_L and I_L are the voltage and current across the load resistance. By eliminating the reverse voltage V^- from Equations (5.1) and (5.2), we obtain

$$2V^+ = V_L + I_L Z_0 \tag{5.3}$$

Equation (5.3) can be viewed as the fundamental equation describing the transmission line voltage and current at the load end. In other words, Equation (5.3) provides the voltage–current characteristic for the load end of the line.

A general Thevenin equivalent circuit has a voltage–current characteristic, for an arbitrary load resistance, of the form

$$V_{eq} = V_L + I_L Z_{eq} \tag{5.4}$$

where V_{eq} is the Thevenin equivalent (open-circuit) voltage and Z_{eq} is the Thevenin equivalent impedance. By comparing Equations (5.3) and (5.4), we conclude that the Thevenin equivalent circuit could be used to replace the line, while producing the same voltage and current at the load, as long as

$$V_{eq} = 2V^+ \tag{5.5}$$

$$Z_{eq} = Z_0 \tag{5.6}$$

Therefore, Equations (5.5) and (5.6) are the Thevenin equivalent parameters for the load end of the line.

At the generator end of a transmission line of characteristic impedance Z_0, with a DC voltage source in series with some generator resistance, the forward and reverse waves on the line satisfy the equations

$$V^+ + V^- = V_{in} \tag{5.7}$$

$$\frac{V^+}{Z_0} - \frac{V^-}{Z_0} = I_{in} \tag{5.8}$$

where V_{in} and I_{in} are the voltage and current, respectively, across the input end of the line. By eliminating the forward voltage V^+ from Equations (5.7) and (5.8), we obtain

$$2V^- = V_{in} - I_{in} Z_0 \tag{5.9}$$

Equation (5.9) can be viewed as the fundamental equation describing the transmission line voltage–current characteristic at the generator end.

The Thevenin equivalent circuit voltage–current characteristic in Equation (5.4) can be modified for the generator end of the transmission line (basically by reversing the reference direction of the current) to yield

$$V_{eq} = V_{in} - I_{in} Z_{eq} \tag{5.10}$$

By comparing Equations (5.9) and (5.10), we conclude that the appropriate Thevenin equivalent circuit for the line at the generator end, which produces the same input voltage and current as the line itself, involves the parameters

$$V_{eq} = 2V^- \tag{5.11}$$

$$Z_{eq} = Z_0 \tag{5.12}$$

Although these parameters are obtained here for a DC excitation, the equivalent circuits remain correct if any of the voltages and currents are functions of time, the lines are lossy (phasor treatment with complex Z_0), or the loads are reactive or nonlinear. We note that Norton equivalent circuits can be developed in a similar manner but are not often used in connection with transmission lines.

The Thevenin equivalent circuits are illustrated in Figure 5.1.

Example: A 10-V battery with an internal resistance of 50 Ω is connected to a transmission line with $Z_0 = 75$ Ω. During the time interval in question, a reverse-going wave with a constant $V^- = 5$ V impinges on the generator end of the line. Determine the parameters for a Thevenin equivalent circuit representing the generator end of the line, and determine the total input voltage and current, during this time interval.

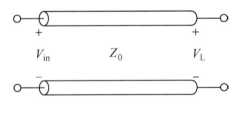

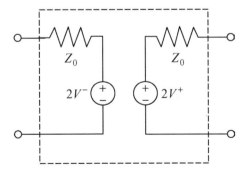

FIGURE 5.1: A length of transmission line and the general Thevenin equivalent circuit that can be used to replace it.

Solution: The Thevenin equivalent parameters are given by

$$V_{eq} = 2V^- = 10 \text{ V}$$

$$Z_{eq} = Z_0 = 75 \text{ } \Omega$$

The total input voltage is the superposition of the battery voltage

$$V_{in}^{battery} = 10\frac{75}{75 + 50} = 6 \text{ V}$$

and the voltage produced by the equivalent Thevenin source

$$V_{in}^{Thev} = 10\frac{50}{50 + 75} = 4 \text{ V}$$

Thus, V_{in} = 10 V. The total input current is the superposition of the battery current

$$I_{in}^{battery} = \frac{V_{in}^{battery}}{75} = 0.08 \text{ A}$$

and the current produced by the Thevenin equivalent circuit

$$I_{in}^{Thev} = -\frac{V_{in}^{Thev}}{50} = -0.08 \text{ A}$$

Therefore, the total input current during this time interval is zero (I_{in} = 0).

5.2   CASCADED TRANSMISSION LINES

It is often the case that a transmission line is cascaded with another line having different character-istic impedance (an example would be traces of different width on a printed circuit board). To study the effect of a sudden change in impedance, consider Figure 5.2, which shows such a system driven by a DC source of 2.0 V. The analysis of a cascaded system begins in the usual fashion, with the determination of various reflection coefficients and the incident voltage on line 1. In this case, Γ_G = 0, Γ_L = −0.333, and the incident voltage is 1.0 V.

At the junction, we can also define a reflection coefficient as experienced by the forward-going wave on line 1 using the notion that the second line, initially uncharged, can be replaced by its Thevenin equivalent resistance Z_{02}. Therefore, we introduce a reflection coefficient

$$\Gamma_F = \frac{Z_{02} - Z_{01}}{Z_{02} + Z_{01}} = 0.333 \qquad (5.13)$$

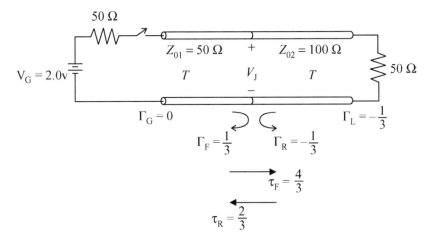

FIGURE 5.2: A transmission line system involving cascaded lines of different Z_0.

representing the impedance mismatch seen by a forward-going wave on line 1 and a second reflection coefficient

$$\Gamma_R = \frac{Z_{01} - Z_{02}}{Z_{01} + Z_{02}} = -0.333 \qquad (5.14)$$

representing the mismatch seen by a reverse-going wave on line 2. These reflection coefficients can be used in a manner identical to that for reflection coefficients at the load or generator ends of the line, in other words $V^{\text{reflected}} = \Gamma V^{\text{incident}}$.

At time $t = T$, when the leading edge of the incident waveform reaches the junction, a reflected voltage is created in line 1 with voltage

$$V_1^- = \Gamma_F V_1^+ = 0.333 \text{ V} \qquad (5.15)$$

At the same time, a second waveform is launched on line 2. Since the lines are directly connected together and since the total voltage appearing across the line at the junction is given by $V_J = V_1^+ + V_1^-$, it follows that

$$V_2^+ = V_J = V_1^+ + V_1^- = (1 + \Gamma_F)V_1^+ \qquad (5.16)$$

It is convenient to define a transmission coefficient

$$\tau_F = \frac{V_2^+}{V_1^+} = (1 + \Gamma_F) = 1.333 \qquad (5.17)$$

so that we can write

$$V_2^+ = \tau_F V_1^+ = 1.333 \text{ V} \tag{5.18}$$

From Equations (5.13) and (5.17), we note that the transmission coefficient can be obtained directly as

$$\tau_F = \frac{2Z_{02}}{Z_{02} + Z_{01}} \tag{5.19}$$

In a similar manner, we can define a transmission coefficient representing the wave launched in line 1 by a negative-going wave in line 2, using

$$\tau_R = \frac{V_1^-}{V_2^-} = (1 + \Gamma_R) = 0.667 \tag{5.20}$$

An alternate way of treating the junction is by using Thevenin equivalent circuits. Figure 5.3 shows the equivalent circuit for the junction voltage during the interval $T < t < 3T$. We see immediately that

$$V_J = \frac{2V_1^+ Z_{02}}{Z_{02} + Z_{01}} = 1.333 \text{ V} \tag{5.21}$$

which is consistent with Equation (5.18) and gives us an alternate way of obtaining the transmission coefficients.

Once we identify the basic parameters, we are able to work the problem using a reflection diagram (Figure 5.4). In this example, reflection and transmission coefficients are used to determine the incremental changes in the transient signal. As $t \to \infty$, both lines charge to a voltage of 1.0 V. In this example, there are no reflections from the generator end of the line, so there is no need to consider the situation where two signals simultaneously impinge on the junction from different directions. When that situation does arise, the resulting voltages are the superposition of the contributions due to the two incident signals.

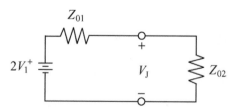

FIGURE 5.3: Equivalent circuit for the junction during the interval $T < t < 3T$.

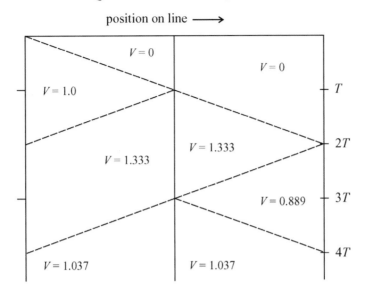

FIGURE 5.4: The reflection diagram for the system in Figure 5.2.

In the previous example, the cascaded lines were directly connected. Figure 5.5 illustrates two lines connected using resistors. If resistors or other components are present between the cascaded lines, the transmission coefficient must be obtained by circuit analysis at the junction to account for the voltage drop across the series components. In this situation, the simple relation presented in Equation (5.17) no longer holds. For the example in Figure 5.5,

$$V_{J2} = V_{J1} \frac{R_P}{R_1 + R_P} \tag{5.22}$$

where R_P is the parallel combination of R_2 and Z_{02},

$$R_P = \frac{R_2 Z_{02}}{R_2 + Z_{02}} \tag{5.23}$$

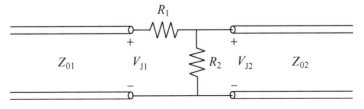

FIGURE 5.5: A junction containing resistors between the two lines.

and the junction voltage is $V_{J1} = V_1^+ (1 + \Gamma_F)$. Consequently, the overall transmission coefficient is given by

$$\tau_F = \frac{V_{J2}}{V_1^+} = (1 + \Gamma_F)\frac{R_P}{R_1 + R_P} \qquad (5.24)$$

The transmission coefficient τ_R may be found from a similar analysis.

Figure 5.6 illustrates another form of cascaded transmission lines within a system where a signal is routed to multiple gates. The matching problem is more difficult in this situation, since a signal will not arrive at the receiver gates at identical times due to the additional path delay to the second gate. Figure 5.7 shows the reflection diagram for the unterminated system if the receiver gates are assumed to have infinite input impedances. The cascaded line is terminated in an open circuit, and the gate voltages never stabilize. A proper termination scheme in this case is illustrated by the parallel match and associated reflection diagram in Figure 5.8. How is the performance of this matching scheme different from that of a series resistor placed at the generator? Would a series or parallel resistor at the junction provide an effective match? (The reader should work out the reflection diagrams for both situations.)

Figure 5.9 shows a fan-out, where one transmission line is split into two. This situation arises whenever a signal needs to be routed to multiple receivers, and could represent a trace driving several gates or a CATV cable carelessly tapped into another. As seen from line 1, the parallel combination of lines 2 and 3 presents an equivalent resistive load equal to $Z_0/2$, with an associated reflection coefficient of $\Gamma_J = -0.333$. Because of this impedance mismatch, some of the signal traveling down line 1 will be reflected at the junction.

Given that the incident wave has voltage V_1^+, the combination of incident and reflected waves in line 1 will add to produce

$$V_J = (1 + \Gamma_J)V_1^+ = 0.667\, V_1^+ \qquad (5.25)$$

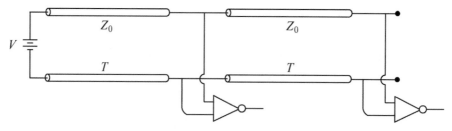

FIGURE 5.6: A signal routed to multiple gates, without proper terminations.

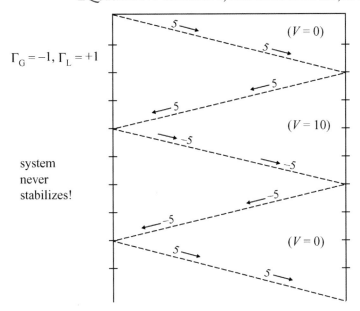

$\Gamma_G = -1, \Gamma_L = +1$

$(V = 0)$

$(V = 10)$

system
never
stabilizes!

$(V = 0)$

FIGURE 5.7: The reflection diagram for the system in Figure 5.6.

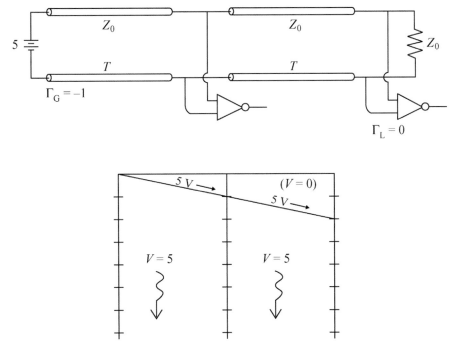

FIGURE 5.8: The system of Figure 5.6 with a proper parallel termination resistor, and the resulting reflection diagram.

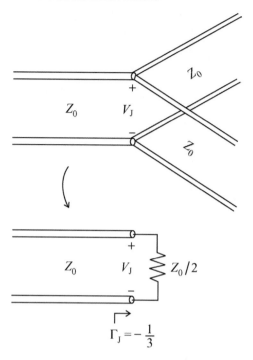

FIGURE 5.9: A transmission line fanning out into two lines at a junction, and the equivalent circuit seen by the initial wave.

The voltage induced in either line 2 or line 3 is equal to V_J, therefore

$$V_2^+ = V_3^+ = 0.667 \, V_1^+ \tag{5.26}$$

The reflected signal carries power

$$P_1^- = -\frac{(V_1^-)^2}{Z_{01}} = -0.111 P_1^+ \tag{5.27}$$

while the signals transmitted into lines 2 and 3 carry power

$$P_2^+ = \frac{(V_2^+)^2}{Z_{02}} = 0.444 P_1^+ \tag{5.28}$$

$$P_3^+ = \frac{(V_3^+)^2}{Z_{03}} = 0.444P_1^+ \qquad (5.29)$$

Thus, tapping into a transmission line in this manner results in an impedance mismatch that in turn causes a reflection of about 11% of the incident power; the remaining power is divided equally between lines 2 and 3.

We observe that the amount of power reflected in this example is proportional to the square of the voltage reflection coefficient, as is easily verified from the formulas:

$$P_1^- = -\left(\frac{V_1^-}{V_1^+}\right)^2 P_1^+ = -(\Gamma_J)^2 P_1^+ \qquad (5.30)$$

This is a general result that should be committed to memory.

Figure 5.10 shows a series fan-out, with an additional imbalance in the characteristic impedances of lines 2 and 3. Series connections are less practical than parallel connections for most transmission lines, and cannot be implemented at all with some forms of line (such as microstrip).

The example in Figure 5.10 can be worked with the aid of Thevenin equivalent circuits at the junction, as illustrated in Figure 5.11.

In this lecture, we considered several examples involving cascaded lines and lines that fan-out from a single line. These applications frequently arise in practical circuit layouts, and introduce mismatches in impedance that can cause significant reflections if not properly designed.

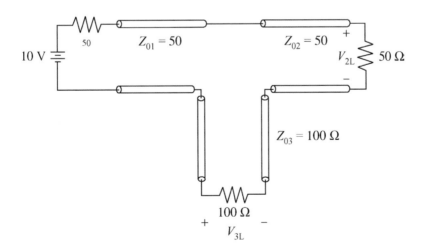

FIGURE 5.10: A series fan-out.

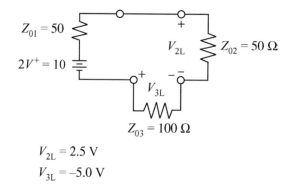

$Z_{01} = 50$

$2V^+ = 10$

V_{2L}

$Z_{02} = 50\ \Omega$

V_{3L}

$Z_{03} = 100\ \Omega$

$V_{2L} = 2.5$ V

$V_{3L} = -5.0$ V

FIGURE 5.11: The equivalent circuit representation and solution for the system in Figure 5.10.

PROBLEMS

5.1 *Fill in the blank:* If a lossless transmission line with characteristic impedance Z_0 has N transmission lines connected to it in a series fan-out, the impedance of each fan-out line must be _____ to maintain an impedance match at the load end of the primary line.

5.2 *Fill in the blank:* If a transmission line fans out to N identical lines in parallel, each having the same characteristic impedance Z_0 as the primary line, the reflection coefficient at the end of the primary line is _____.

5.3 *Fill in the blank:* If a transmission line fans out to N identical lines in series, each having the same characteristic impedance Z_0 as the primary line, the reflection coefficient at the end of the primary line is _____.

5.4 The cascaded transmission line system below has $V_S = 10$ V, $R_G = 75\ \Omega$, $Z_{01} = 75\ \Omega$, $Z_{02} = 50\ \Omega$, and $R_L = 150\ \Omega$. The line is completely discharged for $t \leq 0$.

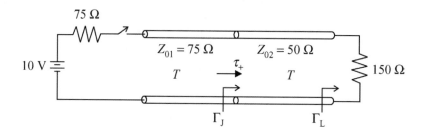

75 Ω

10 V

$Z_{01} = 75\ \Omega$

T

τ_+

$Z_{02} = 50\ \Omega$

T

150 Ω

Γ_J

Γ_L

The switch is closed at $t = 0$. Generate a reflection diagram for the system, and plot the voltage across R_L as a function of time during the interval $0 < t < 8T$, clearly labeling all voltage levels.

5.5 The cascaded transmission line system below has $V_0 = 3.5$ V, $R_G = 25$ Ω, $Z_{01} = Z_{02} = 50$ Ω, and $R_L = 75$ Ω. The lines are connected through a 50-Ω resistor as shown. The line is completely discharged for $t \leq 0$.

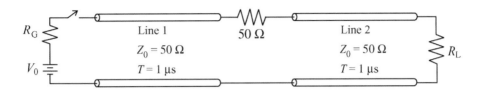

The switch is closed at $t = 0$. Generate a reflection diagram for the system, and plot the voltage across R_L as a function of time during the interval $0 < t < 8T$, clearly labeling all voltage levels.

5.6 Suppose the transmission line system from Problem 5.5 is used with an unknown V_0, R_G, and R_L and excited in such a way that a signal carrying 50 mW of power is launched in line 1 immediately after the switch is closed at $t = 0$.

(a) How much power is transmitted into line 2 during the time interval 1 μs $< t < 3$ μs?

(b) If 33.3% of this power is reflected off the load resistor R_L, what is R_L?

5.7 The cascaded transmission line system below has $V_G = 10$ V, $R_G = 75$ Ω, $Z_{01} = 75$ Ω, $Z_{02} = 50$ Ω, and $R_L = 50$ Ω. The lines are connected through two resistors as shown. The lines each have time delay T and are completely discharged for $t \leq 0$.

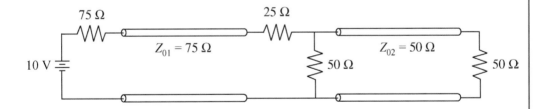

The battery is connected to the system at $t = 0$. Generate a reflection diagram for the system, and plot the voltage across the load resistor as a function of time during the interval $0 < t < 4T$, clearly labeling all voltage levels. How much power is dissipated in the two resistors at the junction?

5.8 You desire to concatenate two transmission lines with differing impedance, as shown below. Design a matching circuit between the two lines that consists of one 10-Ω, one 45-Ω, and one 60-Ω resistor. Provide a sketch of the circuit, and demonstrate that the transmission lines are matched from both directions.

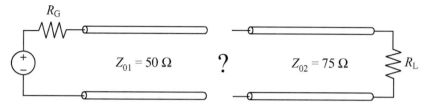

5.9 For the parallel fan-out system shown in the figure below, each line has a one-way transit time delay of T. The lines are uncharged until the switch closes at $t = 0$. Plot the voltages V1, V2, and V3 during the time interval $0 < t < 7T$.

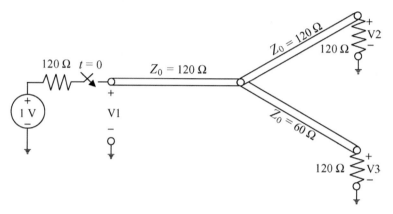

5.10 A microstrip circuit board with substrate thickness of h contains a fan-out that connects the output of a logic gate on chip 0 to the input logic gates of two other chips, as illustrated in the figure below:

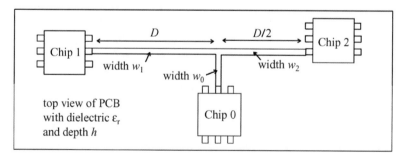

The values of h and ε_r are fixed board parameters, and w_0 is also fixed for this problem. You may use the following approximations for the characteristic impedance and velocity of propagation on microstrip:

$$Z_0 \cong \frac{377}{\sqrt{\varepsilon_r}\left[\dfrac{w}{h} + 2\right]}$$

$$v_p \cong \frac{3 \times 10^8}{\sqrt{\dfrac{\varepsilon_r + 1}{2} + \dfrac{\varepsilon_r - 1}{2} \dfrac{1}{\sqrt{1 + 12h/w}}}}$$

(a) Solve for the two trace widths w_1 and w_2 in terms of the parameters h, ε_r, and w_0 in order to impedance match the circuit at the fan-out junction looking out of chip 0. (Ignore impedance discontinuities at the input of chips 1 and 2.)

(b) Describe how you might design the widths w_1 and w_2 to obtain an impedance match looking out of chip 0 while simultaneously adjusting the propagation velocities so that signals from chip 0 arrive at the exact same time at chips 1 and 2. You do not have to solve for w_1 and w_2, but you should specify the conditions that Z_0 and v_p must satisfy.

CHAPTER 6

Initially Charged Transmission Lines

Objectives: Consider disturbances on transmission lines that are initially charged. Develop a procedure for analyzing charged lines, and use examples to demonstrate the discharging process.

Our previous lectures have considered unenergized lines and focused on the mechanism by which they charged to a steady-state voltage. In this lecture, we consider the reverse situation: lines that are being discharged. The starting point for analyzing charged lines is the steady-state situation considered in previous lectures. In the steady-state situation, forward- and reverse-going waves superimpose to yield a constant voltage and constant current along a transmission line. Such a line is said to be "charged" to that voltage, in a manner similar to a capacitor being charged.

6.1 STEADY-STATE WAVE AMPLITUDES

Figure 6.1 shows a transmission line system. After the switch is closed, the battery creates an initial forward-going wave with voltage $V^+ = 8$ V. This wave propagates toward the load, where it encounters a reflection coefficient of $\Gamma_L = -0.333$. The reflected wave with $V^- = -2.667$ V propagates back toward the generator end of the line, where $\Gamma_G = 0.2$. Forward- and reverse-going waves eventually superimpose to yield a steady-state voltage (see the equivalent circuit in Figure 3.9) of

$$V_{ss} = 20 \left(\frac{25}{75 + 25} \right) = 5.0 \text{ V} \qquad (6.1)$$

and a steady-state current of

$$I_{ss} = \frac{V_{ss}}{R_L} = 0.2 \text{ A} \qquad (6.2)$$

The steady-state forward and reverse waves that combine to produce Equations (6.1) and (6.2) can be determined using the basic solutions from Chapter 3, given by

$$V_{ss} = V_{ss}^+ + V_{ss}^- \qquad (6.3)$$

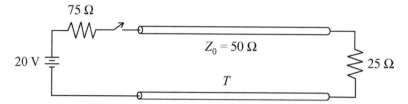

FIGURE 6.1: A transmission line system.

$$I_{ss} = \frac{V_{ss}^+}{Z_0} - \frac{V_{ss}^-}{Z_0} \qquad (6.4)$$

From Equations (6.3) and (6.4), we obtain the equations

$$V_{ss}^+ + V_{ss}^- = 5 \qquad (6.5)$$

$$V_{ss}^+ - V_{ss}^- = I_{ss}Z_0 = 10 \qquad (6.6)$$

and readily determine that $V_{ss}^+ = 7.5$ V, $V_{ss}^- = -2.5$ V, $I_{ss}^+ = 0.15$ A, and $I_{ss}^- = 0.05$ A. These are the total voltage and current wave amplitudes that exist on the system of Figure 6.1 as $t \to \infty$.

We next consider what happens if the transmission line system is perturbed from its steady-state situation. Figure 6.2 shows the transmission line system of Figure 6.1, but with an additional resistor that can be switched into the system. Suppose the system is operated for a long time with the switch in position A, so that the voltage and current waves have reached their steady-state values. (Those are the values determined in the previous paragraph.) At time $t = 0$, the switch is moved from position A to position B. This change disconnects the battery from the line and changes the load resistor at the generator end. We wish to create a reflection diagram showing the state of the system for $0 < t < 4$T.

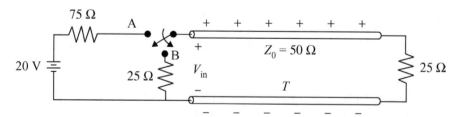

FIGURE 6.2: The system of Figure 6.1 is charged for a long time. At $t = 0$, the switch is moved from position A to position B.

This problem can be approached two ways, one involving the use of Thevenin equivalent circuits and another through consideration of reflection coefficients.

6.1.1   Approach 1: Thevenin Equivalent Circuit Analysis

As determined above, the line voltage at the generator end of the line *before* the switch is moved from position A is $V_{in} = 5$ V. A Thevenin equivalent circuit can be used to determine the total line voltage *after* the switch is moved to position B. An equivalent circuit for the generator end of the line, for $t = 0^+$, is shown in Figure 6.3. This equivalent circuit uses the characteristic impedance $Z_0 = 50$ Ω and has an equivalent voltage $2V_{ss}^- = -5.0$ V. From the equivalent circuit, the new input voltage is $V_{in} = -1.667$ V.

Thus, the disturbance of moving the switch from position A to B changed the total input line voltage V_{in} from $+5$ V to -1.667 V at $t = 0$, a change of -6.667 V. We interpret this result as if the act of closing the switch at $t = 0$ created a new -6.667 V reflection from the input end of the line or, equivalently, a new $V^+ = -6.667$ that must be superimposed with the previous forward-going waves.

Since the transmission line configuration does not change after $t = 0$, with the load reflection coefficient fixed at $\Gamma_L = -0.333$ and the generator reflection coefficient fixed at the new value of $\Gamma_G = -0.333$, the effect of the disturbance can be determined in the usual way. Figure 6.4 shows a reflection diagram depicting the line voltages until $t = 4T$. It is apparent that the line voltage decreases in magnitude, as we should expect since the battery has been disconnected from the line and the system is gradually discharging.

6.1.2   Approach 2: Reflection Coefficient Analysis

Alternatively, we can view the previous example as one in which at $t = 0$ (a) the source is removed and (b) the reflection coefficient is changed. For $t < 0$, the source had contributed a positive-going voltage wave with $V_{source}^+ = 8$ V and the reflection off the 75-Ω generator resistor ($\Gamma_G = 0.2$) had

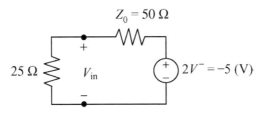

FIGURE 6.3: Thevenin equivalent circuit for the generator end of the transmission line in Figure 6.2, for $t = 0^+$.

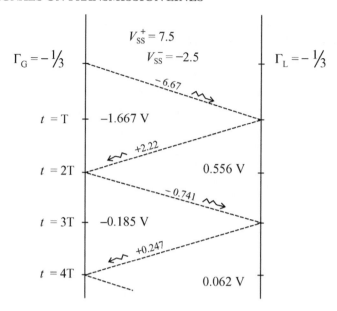

FIGURE 6.4: The reflection diagram illustrating the discharge process.

produced a steady-state voltage wave with $V_{ref}^{+} = \Gamma_G V_{ss}^{-} = -0.5$ V. The combination of these two signals produces the steady-state positive-going wave voltage of 7.5 V. For $t > 0$, the source contributes $V_{source}^{+} = 0$ V; the reflection off the new 25-Ω generator resistor ($\Gamma_G = -0.333$), for $0 < t < 2T$, produces $V_{ref}^{+} = \Gamma_G V_{ss}^{-} = 0.833$ V. This represents a change of -6.667 V in the total positive-going voltage, the same differential determined by the Thevenin equivalent circuit approach.

6.2   A SECOND EXAMPLE

Figure 6.5 shows a transmission line in steady state, charged to 10.0 V. At time $t = 0$, the switch at the load end of the line is closed, changing the load resistance from an open circuit to 25 Ω.

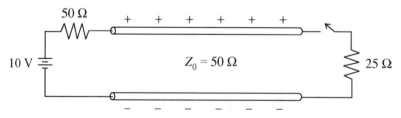

FIGURE 6.5: A transmission line charged to +10.0 V.

Equivalently, the reflection coefficient at the load changes at time $t = 0$ from $\Gamma_L = +1$ to $\Gamma_L = -0.333$.

As in the preceding example, we first decompose the line voltage for $t < 0$ into forward- and reverse-going waves. Using Equations (6.3) and (6.4), we obtain

$$V_{ss}^+ + V_{ss}^- = 10 \tag{6.7}$$

$$V_{ss}^+ - V_{ss}^- = 0 \tag{6.8}$$

and determine that

$$V_{ss}^+ = 5.0 \text{ V} \tag{6.9}$$

$$V_{ss}^- = 5.0 \text{ V} \tag{6.10}$$

(For this example, these values can be determined by inspection, since the line is matched at the generator end. The steady-state line voltage consists of the incident wave and the single reflected wave.)

Once we have obtained the steady-state forward and reverse voltages at $t = 0^-$, we can construct a Thevenin equivalent circuit for the load end of the line at $t = 0^+$ (Figure 6.6). From the equivalent circuit, we find that the load voltage changes from 10.0 to 3.333 V at $t = 0$. We interpret this result as if the act of closing the switch at $t = 0$ created an additional reflection of -6.667 V, which must be superimposed with the previous V_{ss}^- of 5.0 V. Thus, the new total reverse-going wave at time $t = 0^+$ is $V^- = -1.667$ V. Since the line is matched at the generator end, the disturbance will travel back to the generator and then cease, leaving the line in a steady-state situation with $V_{ss} = 3.333$ V. A reflection diagram is given in Figure 6.7.

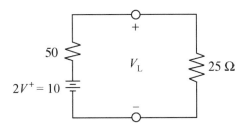

FIGURE 6.6: The Thevenin equivalent circuit for the load end of the line in Figure 6.5 at time at $t = 0^+$.

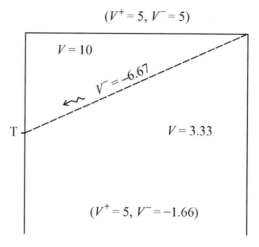

$(V^+ = 5, V^- = 5)$

$V = 10$

$V^- = -6.67$

T

$V = 3.33$

$(V^+ = 5, V^- = -1.66)$

FIGURE 6.7: A reflection diagram for the discharge of the transmission line in Figure 6.5.

Alternatively, we can consider the effect of changing the reflection coefficient at $t = 0$ from $\Gamma_L = 1$ to $\Gamma_L = -0.333$. As the switch is closed, the previous reflection of $(+1)(5.0) = 5.0$ V is changed to a new reflected voltage of $(-0.333)(5.0) = -1.667$ V. On a reflection diagram, the perturbation appears as an incremental reflected voltage of -6.667 V, consistent with the result of the Thevenin equivalent circuit analysis.

6.3   EXAMPLE: DISCHARGE OF LOGIC GATE

A similar situation arises if we consider the transition of a logic gate from a high state to a low state, for the system illustrated in Figure 6.8. Observe that this configuration, studied previously in Figure 4.9, includes a series matching resistor. As the transmission line originally transitioned to a charged state, the initial voltage waves were given by

$$V^+ = 2.5 \text{ V} \tag{6.11}$$

$$V^- = \Gamma_L V^+ = (0.951)(2.5) = 2.378 \text{ V} \tag{6.12}$$

Because the line is series matched, these are also the steady-state forward- and reverse-going voltages. As the driver gate switches to the "off" state at $t = 0$, the equivalent DC source at the generator end of the line changes from 5.0 to 0.0 V, removing the forward-going wave of 2.5 V. Consequently, the effect is as if an incremental wave of -2.5 V is launched down the line in the positive direc-

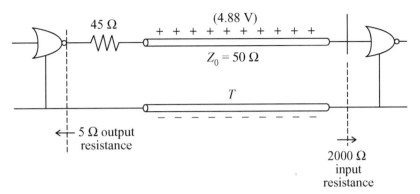

FIGURE 6.8: Two logic gates connected by a transmission line. The line between the gates is charged to a steady-state voltage of 4.88 V.

tion, which subsequently reflects from the impedance mismatch at the load end (Figure 6.9). The matching resistor at the generator end absorbs the reflected transient, and the entire line has been discharged to zero voltage by $t = 2T$.

It is worth noting that, during the transition to a steady-state line voltage of zero, the charge originally stored in the transmission line has drained off through the driver. The current into the

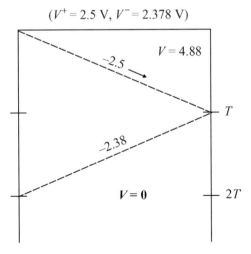

FIGURE 6.9: A reflection diagram for the discharge process for the system shown in Figure 6.8.

line during the steady-state period prior to $t = 0$ is $I_{in} = 2.4$ mA; during the time interval $0 < t < 2T$, the input current is given by

$$I_{in} = -47.6 \text{ mA} \tag{6.13}$$

After time $t = 2T$, the input voltage and current are zero. The process of discharging the transmission line is somewhat analogous to discharging a capacitor, with the time duration of the process dependent on the line length.

6.4 EXAMPLE: CASCADED LINES

A fourth example of an initially charged system is illustrated in Figure 6.10, which involves cascaded transmission lines that have been charged by an external battery (now removed from the system). The switch in the system is closed at $t = 0$ and begins the process of discharging the lines.

Suppose that the lines are charged to 7.5 V. To analyze the discharge process, we first identify the various reflection and transmission coefficients. These are given by

$$\Gamma_G = 1.0 \tag{6.14}$$

$$\Gamma_L = 1.0 \tag{6.15}$$

$$\Gamma_F = -0.333 \tag{6.16}$$

$$\Gamma_R = -0.333 \tag{6.17}$$

$$\tau_F = (1 + \Gamma_F) = 0.667 \tag{6.18}$$

$$\tau_R = (1 + \Gamma_R) = 0.667 \tag{6.19}$$

We also identify the total forward- and reverse-going voltage waves using Equations (6.3) and (6.4). In this situation, $V_{ss} = 7.5$ V and $I_{ss} = 0$ A, so

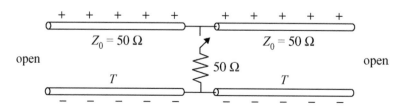

FIGURE 6.10: A cascaded arrangement of transmission lines, charged to a steady-state voltage.

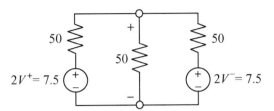

FIGURE 6.11: A Thevenin equivalent circuit for the junction between the transmission lines of Figure 6.10, for $0 < t < 2T$.

$$V_{ss}^{+} = 3.75 \text{ V} \tag{6.20}$$

$$V_{ss}^{-} = 3.75 \text{ V} \tag{6.21}$$

With this information, a Thevenin equivalent circuit for the junction at time $t = 0^{+}$ is developed (Figure 6.11). The junction voltage changes from 7.5 to 5.0 V, indicating that an incremental voltage waveform of -2.5 V is created at the junction and propagates in both directions. The entire reflection diagram is shown in Figure 6.12.

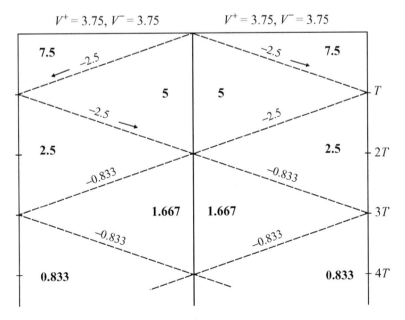

FIGURE 6.12: A reflection diagram for the discharge process.

In this section, we considered the treatment of transients on transmission lines that are initially charged. This is a practical situation, since the process of discharging a line occurs as often as the process of charging a line. In Chapters 8 and 9, we will consider reactive and nonlinear loads, which afford more accurate models of transmission lines arising in the digital situation. In Chapter 7, we investigate finite-duration signals on resistively loaded lines.

PROBLEMS

6.1 As a practical joke, you charge up a long, skinny parallel-plate capacitor to 200 V and leave it in your roommate's sock drawer. Later that day, your roommate touches the end of the charged capacitor with his 100-Ω finger, which happens to be perfectly matched to the transmission line model of these parallel plates.

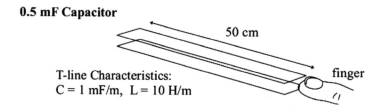

0.5 mF Capacitor

50 cm

T-line Characteristics:
C = 1 mF/m, L = 10 H/m

finger

How long does the discharge last? Plot the voltage along the length of the capacitor at times $t = 0$, $t = T/2$, $t = T$, $t = 3T/2$, and $t = 2T$.

6.2 In the system shown below, the line is initially uncharged. At $t = 0$, the switch is closed.

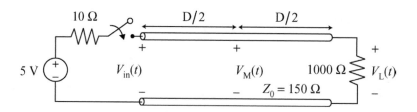

(a) Plot the midline voltage $V_M(t)$ for $0 < t < 6T$, clearly labeling all voltage levels.

(b) After the switch at the generator end of the line has been closed for a long time, it is opened at $t = t_0$. Plot the load voltage $V_L(t)$ for $t_0 < t < (t_0 + 6T)$, clearly labeling all voltage levels.

6.3 In the system shown below, the switch is kept in position A for a long time. At $t = 0$, the switch is moved to position B.

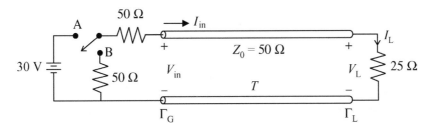

Develop a reflection diagram for the discharge process resulting from the switch changing position, up to $t = 4T$. In addition, plot the load voltage $V_L(t)$ and the input current $I_{in}(t)$ for $0 < t < 4T$.

6.4   In the system shown below, the driver gate can be modeled as a 5-V battery in series with a 10-Ω resistor and a switch, while the receiver gate can be modeled as a 5-kΩ resistor. The transmission line is charged to a steady-state voltage of 4.99 V, representing a "high" logic state. At time $t = 0$, the driver changes to a "low" state, and the 5-V battery is removed from the circuit at the generator end of the line.

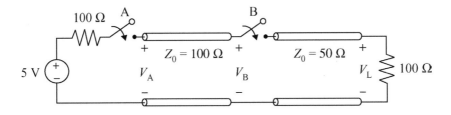

If the receiver logic gate switches to a low state when the applied voltage at its input drops below 1.0 V, at what time t will the receiver gate switch? Develop a reflection diagram for the discharge process and use it to answer the question.

6.5   Consider the transmission line system shown below:

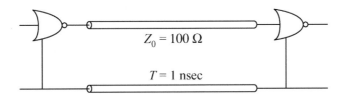

This system undergoes a number of state changes in the sequence described below:
   State 0: Both switches are open and both lines are uncharged
   State 1: Immediately after switch A is closed, with B open

State 2: Switch A has been closed for a long time with B open

State 3: Immediately after switch B is closed, with switch A closed for a long time

State 4: Both switches have been closed for a long time

State 5: Immediately after switch A is opened, with B closed

State 6: Switch A has been open for a long time, with B closed

Based on these states, complete the following table. (Note that V_1^+ is the forward-going wave on line 1, V_1^- is the backward-going wave on line 1, etc.)

	V_A	V_B	V_L	V_1^+	V_1^-	V_2^+	V_2^-
State 0	0	0	0	0	0	0	0
State 1							
State 2							
State 3							
State 4							
State 5							
State 6	0	0	0	0	0	0	0

6.6   Consider the transmission line system shown below:

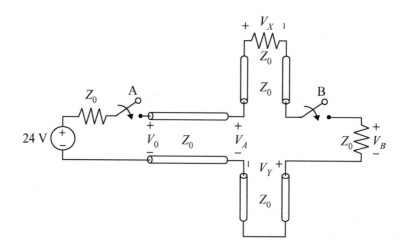

This system is switched in the sequence described below:

   State 0: Both switches are open and all lines are uncharged.
   State 1: Immediately after switch A is closed.
   State 2: Switch A has been closed for a long time.
   State 3: Immediately after switch B is closed.
   State 4: Both switches have been closed for a long time.

Based on these states, complete the following table:

	V_0	V_A	V_B	V_X	V_Y
State 0	0	0	0	0	0
State 1					
State 2					
State 3					
State 4					

6.7   Consider the transmission line system shown below:

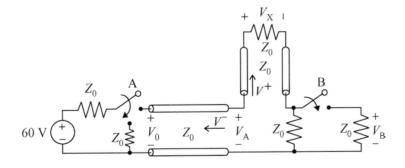

This system is switched in the sequence described below:

   State 0: Both switches are open and all lines are uncharged
   State 1: Immediately after switch A is connected to the first transmission line
   State 2: Switch A has been closed for a long time
   State 3: Immediately after switch B is closed

State 4: Switch B has been closed for a long time

State 5: Immediately after switch A is connected to the dead Z_0 load

Based on these states, complete the following table:

	V_0	V_A	V_B	V_X	V^+	V^-
State 0	0	0	0	0	0	0
State 1						
State 2						
State 3						
State 4						
State 5						

CHAPTER 7

Finite Duration Pulses on Transmission Lines

Objectives: Extend the application of transmission line theory to finite duration pulses excited on a resistively loaded line. Examine the consequences of mismatches for a number of examples.

Up until now, we have considered transmission lines driven by DC sources. In this lecture, we turn our attention to the case of a signal with a short temporal (time) duration. Consider the transmission line system shown in Figure 7.1. The generator in this example produces a single triangular waveform, which is launched down the line. Our task is to determine the load voltage versus t for $0 < t < 4T$.

For conceptual clarity, consider first the situation on the line at time $t = 3T/4$. The leading edge of the waveform will have traveled three fourths of the distance to the load, and the voltage snapshot at this time is presented in Figure 7.2a. Observe that the leading edge of the triangle wave is the edge first produced by the generator (at time $t = 0$), and this leading edge has reached the location $z = 3D/4$ at time $t = 3T/4$. The trailing edge of the signal left the generator at time $t = T/2$ and has only reached the location $z = D/4$. The peak voltage of the signal is 20.0 V, the value determined using a voltage divider at the line input.

Figure 7.2b shows the waveform at time $7T/4$, after the entire signal has reflected from the load end of the line. The mismatch at the load creates a reflection with $\Gamma_L = 0.333$, and the peak amplitude of the reflected signal is therefore 6.667 V, since $V^- = \Gamma_L V^+$. The leading edge has reached the point $z = D/4$, traveling in the $-z$ direction.

Figure 7.2c shows the situation at $t = 11T/4$, after the signal reflects from the generator end of the line, where $\Gamma_G = -0.333$. The peak voltage of the triangle wave is now -2.222 V, based on $V^- = 6.667$ V and $\Gamma_G = -0.333$.

In most respects, a finite-duration waveform can therefore be treated in a similar fashion to that of a DC driver. The primary difference is that the total signal is given by the single waveform

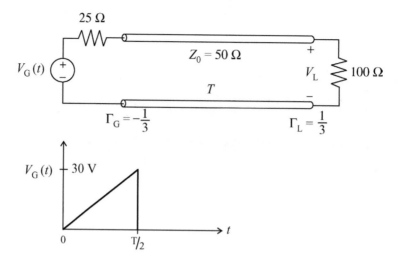

FIGURE 7.1: Transmission line driven by triangular waveform.

shown in Figure 7.2, in contrast to the case of a DC source where the total signal is the sum of the various negative- and positive-going waves. As a consequence of the finite energy carried by the triangle wave in this example, the signal is reduced upon each reflection since some of its power is dissipated by the resistors it encounters at the ends of the line. As $t \to \infty$, the voltage will drop to zero along the line.

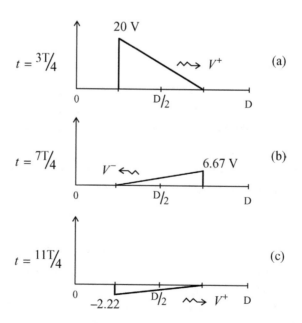

FIGURE 7.2: Triangular waveform distribution along transmission line at different times.

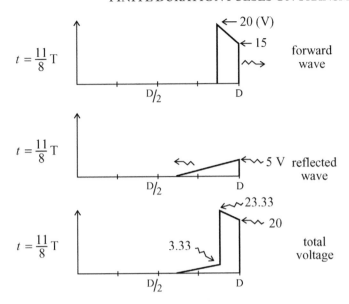

FIGURE 7.3: Voltage on the line at $t = 11T/8$.

We now return to our task of identifying the load voltage for $0 < t < 4T$. The graphs shown in Figure 7.2 avoid the interesting case of a waveform impinging onto the load end of the line. After the leading edge of the waveform reaches the load, a reflection is created, and for a short interval thereafter, the resulting line voltage is the superposition of the reflected part of the waveform and the incident part. For example, the voltage at time $t = 11T/8$ is shown in Figure 7.3, which illustrates the superposition of a negative-going wave (the leading part of the signal) and a positive-going wave (the rest of the signal, which has not yet reflected). Conceptually, this superposition process might be best thought of as the combination of two separate waveforms, each containing a piece of the original signal. The graphs in Figure 7.2 can be used to assist in the construction of these intermediate waveforms.

At any instant, the total voltage across the load end of the line can be found from

$$V_{\mathrm{L}}(t) = V^{+}|_{z=D,t} + V^{-}|_{z=D,t} \qquad (7.1)$$

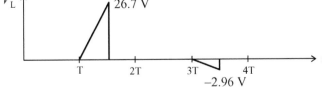

FIGURE 7.4: Total voltage on transmission line, as function of time.

while the voltage across the input end of the line has a similar form

$$V_{in}(t) = V^+|_{z=0,t} + V^-|_{z=0,t}$$
(7.2)

With these ideas, the total load voltage as a function of time is easily constructed, as shown in Figure 7.4. Observe that the peak voltage induced across the load at time $t = 3T/2$ is 26.667 V, the sum of the incident signal and the reflected signal peak voltages.

A reflection diagram can be constructed for finite-duration signals, as shown for the previous example in Figure 7.5. Note that it is necessary to identify the leading and trailing edges of the signal on the diagram, instead of just the leading edge as in the case of a DC source. The reflection diagram clearly shows the regions of space–time where waveform overlap occurs and can be interpreted in the usual manner in other respects.

As a second example, consider Figure 7.6, involving a line driven by a finite duration pulse with a duration longer than time T. This case is somewhat more complicated than the previous example since the forward- and reverse-going waveforms will be superimposed over a larger percentage of the line. The initial positive-going waveform has amplitude of 5 V, which generates a reflected waveform at the load end of the line beginning at time T having amplitude -2.5 V. Figure 7.7 illustrates snapshots of the line voltages at times $1.5T$, $1.75T$, $2.0T$, and $2.25T$. At time $1.5T$, the forward-going waveform occupies the entire line while the leading edge of the reflected wave has reached the halfway point back toward the generator end. At time $1.75T$, the trailing edge of the forward-going waveform coincides with the leading edge of the reflected waveform, producing a snapshot with a resulting voltage of 2.5 V. At time $t = 2T$, the leading edge of the reflected waveform

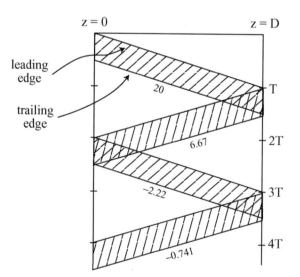

FIGURE 7.5: Reflection diagram for the first example.

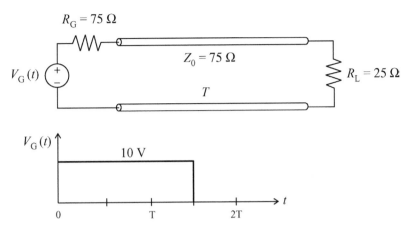

FIGURE 7.6: Second example, where the signal has a duration exceeding T.

has been absorbed by the matched resistive termination at the generator end of the line, while the trailing edge of the +5 V waveform is three-quarters of the way toward the load.

As with any transmission line problem, the total voltage appearing across the load resistor is the superposition of the incident and reflected waveforms. Figure 7.8 shows the load voltage as a

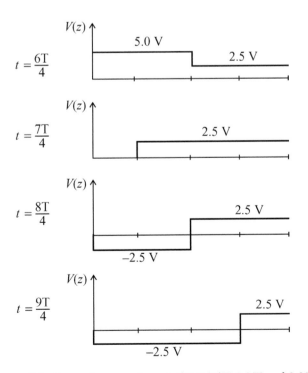

FIGURE 7.7: Snapshots of the line voltages at times $1.5T$, $1.75T$, $2.0T$, and $2.25T$.

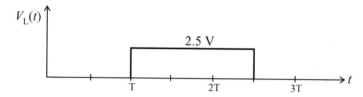

FIGURE 7.8: Load voltage as a function of time.

function of time. A reflection diagram can be constructed for this example; we leave that task to the reader as an exercise!

These two examples demonstrate that the treatment of finite-duration signals on transmission lines can be accomplished in much the same manner as that of DC excitations. Plots such as Figures 7.2 and 7.3 are an invaluable aid to visualizing the behavior of waveforms, especially when forward- and reverse-going parts of the signal are superimposed.

As a final example, suppose we have a transmission line with length D, one-way transit time T, and reflection coefficients of $\Gamma_G = -0.8$ and $\Gamma_L = +0.8$. At $t = 0$, an impulse (Dirac delta function) with amplitude 1.0 enters the source side of the line. We would like to determine the voltage observed at the load end of the line, at the source end of the line, and at the very center of the line ($z = D/2$).

The voltage impulse can be tracked as it reflects from the two ends of the line. The 1.0-V impulse first reaches the load end at time T, where the reflected voltage impulse has amplitude

$$V^- = \Gamma_L V^+ = 0.8 \text{ V} \tag{7.3}$$

Since the actual load voltage is the superposition of the incident and reflected signals, the load voltage is the sum of these: $V_L = 1.8$ V.

The reflected impulse reaches the source end of the line at time $2T$, where it produces the forward-going reflection

$$V^+ = \Gamma_G V^- = -0.64 \tag{7.4}$$

At the generator end of the line at $t = 2T$, the total voltage is the sum of the 0.8-V impulse and the -0.64-V reflection, to produce $V_G = 0.16$ V.

At $t = 3T$, when the forward-going impulse impinges on the load end of the line, a reflection with voltage

$$V^- = \Gamma_L V^+ = -0.512 \tag{7.5}$$

is produced, and the total load voltage impulse has amplitude of $V_L = -1.152$ V.

a. The voltage observed at the transmission line loaded:

b. The voltage observed at the transmission line source:

c. The voltage observed exactly halfway down the transmission line:

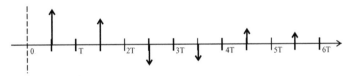

FIGURE 7.9: The voltages observed at (a) the load end, (b) the source end, and (c) the halfway point.

The reflections continue, until the energy is dissipated as $t \to \infty$. Figure 7.9 shows a sketch of the various voltages for $0 < t < 6T$.

PROBLEMS

7.1 The signal shown below in the figure on the left is produced by the generator in the transmission line system on the right. Plot the voltage across the load resistor for $0 < t < 4T$, clearly labeling all important amplitudes.

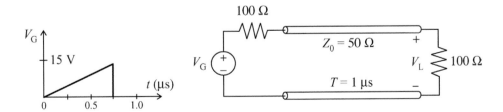

7.2 A transmission line system and generator voltage are shown below. Plot the line voltage and line current as a function of z at $t = 1.5T$, clearly labeling all important amplitudes.

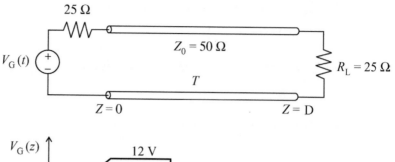

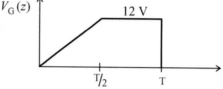

7.3   Refer to the transmission line system and input line voltage plot shown below in order to answer the following questions:

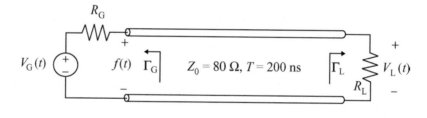

$$V_L(t) = 0.1 \sum_{n=0}^{\infty} (-0.7)^n f\left(t - [2n + 1]T\right)$$

(a)   If $V_L(t)$ is as given immediately above, determine the reflection coefficients Γ_L and Γ_G.

(b)   Determine the values of R_L and R_G.

(c)   If the voltage across the input end of the line, $f(t)$, is a 5.0-V square pulse of 200-ns duration, as shown above, sketch $V_L(t)$ for $0 < t < 1$ μs.

(d)   For the $f(t)$ given in part (c), determine $V_G(t)$.

(e) Now, assume that the same 5-V pulse in the previous questions represents unipolar signaling of bits in a digital system ($1 = 5$ V square wave, $0 = 0$ V signal). If one bit is transmitted every 200 ns, make a plot of the first 4 μs of transmission line output, $V_L(t)$, in response to the 8-bit input sequence 11101101.

7.4 Part of a transmission line system is shown below:

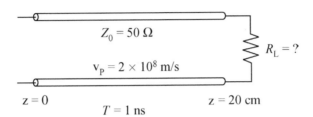

An observer located at $z = 5$ cm sees the following line voltages during the time interval $0 < t < 2$ ns:

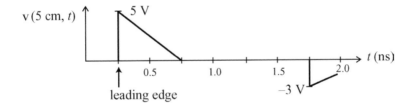

Based on this information, provide answers to the following:
(a) What is the value of R_L?
(b) Sketch the line voltage versus position z at time $t = 1.5$ ns. Label all peak voltages!
(c) Sketch the line current versus z at $t = 1.25$ ns. Label all peak currents!

7.5 Consider an uncharged transmission line with one-way transit time T, total length D, and reflection coefficients $\Gamma_G = 0.5$ and $\Gamma_L = -0.5$. At $t = 0$, an ideal impulse $f(t) = 16\delta(t)$, is applied to the input end of the line. Sketch the following functions for the interval $0 < t < 6T$, clearly labeling all amplitudes and showing the correct modulus (sign) and relative amplitudes in the plots:
(a) The voltage at the load end of the transmission line.
(b) The voltage at the input or source end of the transmission line.
(c) The voltage one quarter of the distance down the transmission line ($z = D/4$).

7.6 The figure below shows an overhead view of microstrip traces linking the output of chip 1 with two other chips on a high-speed digital printed circuit board, consisting of a dielectric substrate and ground plane below the plane shown in the figure. The transit time is labeled on each leg of transmission line. All sources, loads, and fan outs are mismatched. A very short pulse (treat as an impulse) is sent to the output of chip 1 at time $t = 0$. List, in order, the first 10 unique times t_i that the input would be nonzero at chip 2.

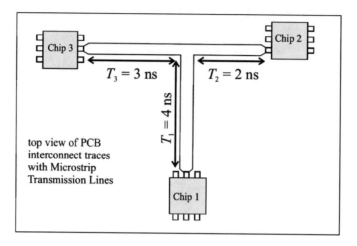

CHAPTER 8

Transmission Lines with Reactive Terminations

Objectives: Extend transmission line theory to account for capacitor or inductor terminations. Review the steps involved in solving differential equations for reactive circuits. Introduce examples illustrating the additional signal distortion possible due to parasitic reactances.

Previous lectures have explored transmission lines terminated with resistive loads. In reality, most line terminations contain some reactance, such as the shunt capacitance arising from the pads of an integrated circuit. For example, Table 8.1 shows typical values of resistance and capacitance associated with several types of digital logic. The purpose of this lecture is to extend transmission line theory in order to incorporate the reactance into the analysis, and to study examples illustrating the resulting signal distortion. The presence of reactance generally provides an additional signal delay that must be accounted for in the design of digital systems, as well as an impedance mismatch that contributes to reflections, signal distortion, and "noise" in general.

We will employ the classical solution of differential equations in order to analyze reactance and omit the use of Laplace transform methods. Students familiar with Laplace transform methods may find it easier to employ that methodology, which is developed in the context of transmission lines by Rosenstark [1]. Two practical problems we consider are transmission lines with RC circuit loads and transmission systems with inductance in the associated power and grounding paths.

One note of clarification, in this chapter, we use L and C to denote the inductance and capacitance of discrete components, not the inductance per unit length and capacitance per unit length of transmission lines, as we did in Chapters 1–2.

Before embarking on a quantitative analysis of reactive loads, we first consider a qualitative study. Figures 8.1a and 8.1b show the load voltage for short- and open-circuit loads, respectively, assuming that the transmission line under consideration is excited with a DC generator and

TABLE 8.1: Approximate range of equivalent input resistance and capacitance for several types of digital logic, as seen from off the chip

LOGIC FAMILY	RESISTANCE	CAPACITANCE (pF)
TTL	$>5\ \Omega$	$\cong 1$–3
ECL	$\cong 50\ k\Omega$	$\cong 1$–3
CMOS	$>10\ M\Omega$	$\cong 1$–3

reflections are eliminated by a suitable series matching resistor at the generator end of the line. The short-circuited line has $v_L(t) = 0$ for all t, while the open-circuited line exhibits a step jump in voltage from 0 to $2V^+$ at time $t = T$.

Now, suppose the line is terminated in a capacitor (initially uncharged). At time $t = T$, when the leading edge of the DC waveform reaches the load, the capacitor will appear as a short circuit across the end of the line, and the behavior of the voltage at the load will be similar to that shown

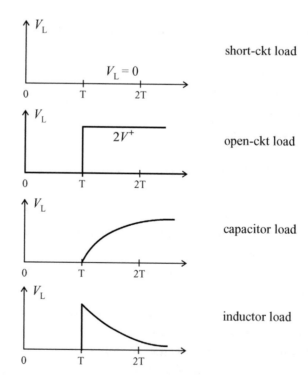

FIGURE 8.1: Qualitative picture of what happens at the load end of the line.

in Figure 8.1a. As the capacitor charges, however, it will eventually look like an open-circuit load and the load voltage will assume the character shown in Figure 8.1b as $t \to \infty$. Thus, qualitatively, we expect to observe a load voltage of the form depicted in Figure 8.1c, consisting of a gradually rising voltage that asymptotically approaches a voltage of $2V^+$ as $t \to \infty$. An important aspect of this behavior, to be studied below, is the time constant associated with the voltage transition.

For a line terminated in an inductor, the initial ($t = T$) situation at the load is similar to that of an open circuit. As the inductor is energized, it eventually assumes the characteristics of a short circuit. Thus, the load voltage at $t = T^+$ should follow the form shown in Figure 8.1b and immediately jump to a voltage of $2V^+$. As the inductor is energized, the voltage gradually drops back to zero as depicted in Figure 8.1d.

Figure 8.2 shows the voltages at the generator end of the transmission line for the four situations depicted in Figure 8.1. With practice, one can determine the type of load on a transmission line by inspecting the reflected signal at the generator end. In fact, measurements made using an instrument known as a *time domain reflectometer* depend on this type of knowledge for their interpretation [2].

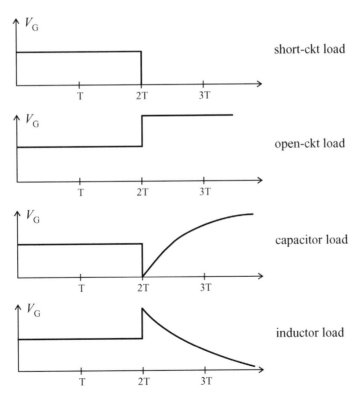

FIGURE 8.2: Qualitative picture of what happens at the input end of the line.

We now embark on a quantitative analysis of reactive terminations. Figure 8.3 shows a transmission line terminated with a capacitance at the load end. The generator end of the line involves a resistive circuit, and therefore, the signal is launched down the line in accordance with previous examples. For the transmission line system shown in Figure 8.3, the waveform initially launched down the line is a DC signal with voltage coefficient $V^+ = V_G/2$. The leading edge of this signal reaches the load end of the line at time $t = T$.

The basic equations for the voltage and current at the load end of the transmission line have described previously in Chapters 3 and 5 and are given by

$$v_L(t) = v^+(t,D) + v^-(t,D) \tag{8.1}$$

$$i_L(t) = \frac{v^+(t,D)}{Z_0} - \frac{v^-(t,D)}{Z_0} \tag{8.2}$$

where $v^+(t,D)$ and $v^-(t,D)$ represent the instantaneous forward- and reverse-going voltage waves at the load end of the line $(z = D)$. Equations (8.1) and (8.2) can be combined to eliminate $v^-(t,D)$, yielding

$$i_L(t) = \frac{2v^+(t,D)}{Z_0} - \frac{v_L(t)}{Z_0} \tag{8.3}$$

Equation (8.3) can be thought of as a single basic equation relating the voltage and current at the load end of the transmission line to the incident voltage waveform. The basic equation for the voltage and current at the capacitor is

$$i_L(t) = C\frac{dv_L}{dt} \tag{8.4}$$

By combining Equations (8.3) and (8.4), we obtain the differential equation

$$\frac{dv_L}{dt} + \frac{1}{Z_0 C}v_L(t) = \frac{2v^+(t,D)}{Z_0 C} \tag{8.5}$$

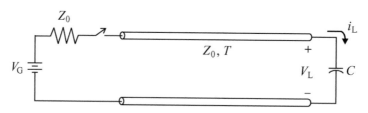

FIGURE 8.3: Transmission line with capacitive load.

For this example, $v^+(t,D) = 0.5V_G u(t - T)$, where $u(t)$ is a unit step function.

The first-order differential equation in Equation (8.5) has a solution that is the combination of a *homogeneous* solution (in the absence of a forcing function) and a *particular* solution (due to the specific forcing function on the right-hand side of the equation). The reader is assumed to have a nodding acquaintance with procedures for solving differential equations, and we will not review these in any detail here. It is sufficient to propose a homogeneous solution of the form

$$v_L(t) = Ae^{-(t-T)/\tau} \tag{8.6}$$

where A and τ are constants to be determined. By substituting Equation (8.6) into the homogeneous equation

$$\frac{dv_L}{dt} + \frac{1}{Z_0 C}v_L(t) = 0 \tag{8.7}$$

we immediately determine that Equation (8.6) satisfies the equation as long as

$$\tau = Z_0 C \tag{8.8}$$

The parameter τ is the *time constant* of the response. The coefficient A will be determined by an initial condition, as described shortly.

For the type of forcing function used in this situation, the particular solution must have the general form

$$v_L(t) = E + Ft \tag{8.9}$$

where E and F are constants. Since conditions at the load end of the transmission line stabilize as the capacitor charges to its steady-state value,

$$\frac{dv_L}{dt} \to 0 \text{ as } t \to \infty \tag{8.10}$$

and

$$v_L(t) \to 2V^+ = V_G \text{ as } t \to \infty \tag{8.11}$$

Therefore $E = V_G$ and $F = 0$ in Equation (8.9).

By combining the homogeneous and particular parts of the solution, we have

$$v_L(t) = 0 \text{ for } t < T \tag{8.12}$$

$$v_L(t) = V_G + Ae^{-(t-T)/\tau} \text{ for } t > T \tag{8.13}$$

where τ is given in Equation (8.8). To determine the remaining coefficient A, we must specify the initial state of the capacitor. In our example, the capacitor and the transmission line are initially uncharged, meaning that

$$v_L(T^+) = 0 \tag{8.14}$$

Therefore, $A = -V_G$. The complete solution for the load voltage is, therefore,

$$v_L(t) = V_G(1 - e^{-(t-T)/\tau})u(t - T) \tag{8.15}$$

Figure 8.4 shows a plot of the voltage. Because of the capacitance, the load voltage gradually transitions to its steady-state value in accordance with the behavior of a typical RC circuit. The current through the load is obtained as

$$i_L(t) = C\frac{dv_L}{dt} = \frac{V_G}{Z_0}e^{-(t-T)/\tau}u(t - T) \tag{8.16}$$

and is also sketched in Figure 8.4.

As a specific example, consider the previous situation with $V_G = 5$ V, $T = 1$ ns, $Z_0 = 50$ Ω, and $C = 10$ pF. The time constant associated with the capacitor is

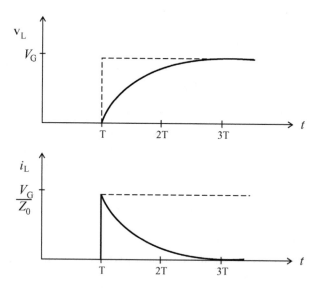

FIGURE 8.4: Load voltage and current on the system shown in Figure 8.3.

$$\tau = Z_0 C = 0.5 \text{ ns} \tag{8.17}$$

The initial waveform reaches the load end of the line at time $t = 1.0$ ns. The load voltage reaches 50% of its steady state value at $t = 1.35$ ns, but does not reach 4.5 V, 90% of its steady-state value, until $t = 2.15$ ns. Thus, the time needed for the capacitor to charge to 90% of its steady-state value in this case exceeds that of the transit time of the line.

As a second example, suppose we use the previous system for the finite-duration generator waveform given in Figure 8.5, which is nonzero for $0 < t < 0.5T$. The initial situation at the load is the same as if the generator were a DC source, so the previous solution in Equation (8.15) suffices for the time $t < 1.5T$. At time $t = 1.5T$, the trailing edge of the positive-going waveform reaches the load end of the line, and the forcing function vanishes. Consequently, for $t > 1.5T$, the situation at the load end of the line is described by just the homogeneous part of the differential equation

$$\frac{dv_L}{dt} + \frac{1}{Z_0 C}v_L(t) = 0 \tag{8.18}$$

subject to the initial condition that $v_L(1.5T) = V_G(1 - e^{-(0.5T)/\tau})$. Thus, the solution for $t > 1.5T$ has the form

$$v_L(t) = Ge^{-(t-1.5T)/\tau} \tag{8.19}$$

where the coefficient G is equal to $v_L(1.5T)$ and τ is given in Equation (8.8).

For the specific numerical values $V_G = 5$ V, $T = 1$ ns, $Z_0 = 50$ Ω, and $C = 10$ pF, the voltage is given by

$$v_L(t) = 0 \qquad (t < 1) \tag{8.20}$$

$$v_L(t) = 5(1 - e^{-(t-1)/0.5}) \qquad (1 < t < 1.5) \tag{8.21}$$

$$v_L(t) \cong 3.16 \, e^{-(t-1.5)/0.5} \qquad (t > 1.5) \tag{8.22}$$

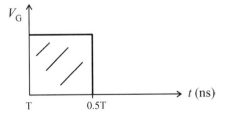

FIGURE 8.5: Finite generator waveform input to system.

where t is in nanoseconds and v_L is in volts. Figure 8.6 shows a plot of this result. Observe that due to the relatively long time required to charge the capacitor in this case, the load voltage only reaches a maximum value of approximately 3.16 V, then decays back to zero. Consequently, if this signal was received by a logic gate in a digital system, the gate would be unlikely to trigger.

The incorporation of reactance allows us to consider other practical problems in transmission line systems, such as the presence of inductance in the power supply and grounding circuitry. Consider Figure 8.7, which depicts two logic gates connected by a transmission line. The driver gate in this case might be located on a chip that is in turn connected to the power supply and ground through a meandering pathway that involves an appreciable amount of inductance. When large numbers of gates attempt to switch state at the same time, the inductance can result in power supply compression on the chip and a phenomenon known as *simultaneous switching noise*. To investigate this situation, suppose we model the system in Figure 8.7 by the equivalent circuit in Figure 8.8, consisting of a 5-V battery driving an open-circuited line. The inductance L appears in series with the battery and the matching resistor.

The basic equation for the transmission line voltages and currents at the generator end is similar to Equation (8.3) and can be written in the form

$$2v^-(t,0) = v_{in}(t) - i_{in}(t)Z_0 \tag{8.23}$$

An equation relating voltage and current within the generator circuit is

$$V_G = L\frac{di_{in}}{dt} + Z_0 i_{in}(t) + v_{in}(t) \tag{8.24}$$

By combining Equations (8.23) and (8.24) to eliminate $v_{in}(t)$, we obtain the differential equation

$$\frac{di_{in}}{dt} + \frac{2Z_0}{L}i_{in}(t) = \frac{V_G - 2v^-(t,0)}{L} \tag{8.25}$$

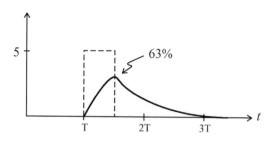

FIGURE 8.6: Plot of result.

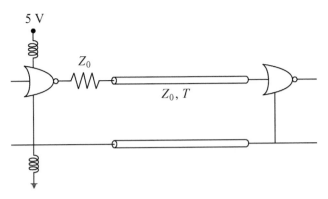

FIGURE 8.7: Two logic gates connected by a transmission line.

Consider the initial time interval $0 < t < 2T$. The homogeneous solution has the form

$$i_{\text{in}}(t) = Ae^{-t/\tau}u(t) \tag{8.26}$$

where the time constant τ is readily determined to be

$$\tau = \frac{L}{2Z_0} \tag{8.27}$$

If the line is initially uncharged, $v^-(t,0) = 0$ for $0 < t < 2T$, and the particular solution can be determined as if the line was infinitely long, using

$$i_{\text{in}}(t) \rightarrow \frac{V_G}{2Z_0} \text{ as } t \rightarrow \infty \tag{8.28}$$

Combining the homogeneous and particular solutions produces

$$i_{\text{in}}(t) = \frac{V_G}{2Z_0} + Ae^{-t/\tau}, \text{ for } 0 < t < 2T \tag{8.29}$$

The remaining coefficient A is determined by the initial condition for an unenergized inductor,

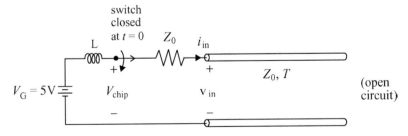

FIGURE 8.8: Equivalent circuit for the system in Figure 8.7.

$$i_{in}(0^+) = 0 \tag{8.30}$$

which yields the solution for input current

$$i_{in}(t) = \frac{V_G}{2Z_0}(1 - e^{-t/\tau}), \quad \text{for } 0 < t < 2T \tag{8.31}$$

By combining this result with Equation (8.23), the input voltage is found as

$$v_{in}(t) = \frac{V_G}{2}(1 - e^{-t/\tau}), \quad \text{for } 0 < t < 2T \tag{8.32}$$

The preceding analysis involved voltages and currents at the generator end of the transmission line. The initial positive-going waveform on the line itself is given for $0 < t < 2T$ by

$$v^+(t,0) = \frac{V_G}{2}(1 - e^{-t/\tau}) \tag{8.33}$$

Since the transmission line in Figure 8.8 is open-circuited, the load voltage takes the form $v_L(t) = (1 + \Gamma_L)v^+(t, D)$, or

$$v_L(t) = V_G(1 - e^{-(t-T)/\tau}), \quad \text{for } T < t < 3T \tag{8.34}$$

The leading edge of the reflection from the load end arrives back at the generator end at time $t = 2T$. From Equation (8.33) and (8.34), we determine that

$$v^-(t,0) = v^+(t - 2T,0) = \frac{V_G}{2}(1 - e^{-(t-2T)/\tau}), \quad \text{for } 2T < t < 4T \tag{8.35}$$

Because the forcing function in the differential equation at time $t = 2T$ changes from a constant to a function of time, the particular solution at time $t = 2T^+$ is more complicated than a simple constant. In this case, the appropriate particular solution for $2T < t < 4T$ is

$$i_{in}(t) = \frac{V_G t}{L}e^{-(t-2T)/\tau} \tag{8.36}$$

(The reader should substitute this result back into Equation (8.25) to verify its validity.) Combining the homogeneous and particular solutions produces

$$i_{in}(t) = Be^{-(t-2T)/\tau} + \frac{V_G t}{L}e^{-(t-2T)/\tau} \tag{8.37}$$

The undetermined coefficient B is found from the "initial" condition

$$i_{in}(2T^+) = i_{in}(2T^-) = \frac{V_G}{2Z_0}(1 - e^{-2T/\tau}) \tag{8.38}$$

which yields the result

$$i_{in}(t) = \frac{V_G}{2Z_0} \left\{ (1 - e^{-2T/\tau}) + \frac{t - 2T}{\tau} \right\} e^{-(t-2T)/\tau}$$

$$= \frac{V_G}{2Z_0} \left\{ e^{-(t-2T)/\tau} - e^{-t/\tau} + \frac{t - 2T}{\tau} e^{-(t-2T)/\tau} \right\}, \quad \text{for } 2T < t < 4T \qquad (8.39)$$

By combining Equations (8.39) and (8.23), we determine that during the interval $2T < t < 4T$

$$v_{in}(t) = \frac{V_G}{2} \left\{ 2 - e^{-(t-2T)/\tau} - e^{-t/\tau} + \frac{t - 2T}{\tau} e^{-(t-2T)/\tau} \right\} \qquad (8.40)$$

Furthermore, using Equation (8.40) and the fact that $v_{in}(t) = v^+(t,0) + v^-(t,0)$, we obtain

$$v^+(t,0) = \frac{V_G}{2} \left\{ 1 - e^{-t/\tau} + \frac{t - 2T}{\tau} e^{-(t-2T)/\tau} \right\}, \quad \text{for } 2T < t < 4T \qquad (8.41)$$

meaning that, for the time interval $3T < t < 5T$

$$v^+(t,L) = \frac{V_G}{2} \left\{ 1 - e^{-(t-T)/\tau} + \frac{t - 3T}{\tau} e^{-(t-3T)/\tau} \right\} \qquad (8.42)$$

and, finally,

$$v_L(t) = V_G \left\{ 1 - e^{-(t-T)/\tau} + \frac{t - 3T}{\tau} e^{-(t-3T)/\tau} \right\}, \quad \text{for } 3T < t < 5T \qquad (8.43)$$

This procedure for developing the solution in each time interval can be continued indefinitely, with the various quantities successively determined as indicated above. (Readers familiar with Laplace transformations might find the approach used by Rosenstark [1] less tedious than the step-by-step solution of the differential equations for each time interval!) Eventually, the entire line charges to a voltage of V_G.

To summarize, we obtained the input voltage

$$v_{in}(t) = 0, \quad \text{for } t < 0$$

$$v_{in}(t) = \frac{V_G}{2}(1 - e^{-t/\tau}), \quad \text{for } 0 < t < 2T$$

$$v_{in}(t) = \frac{V_G}{2} \left\{ 2 - e^{-(t-2T)/\tau} - e^{-t/\tau} + \frac{t - 2T}{\tau} e^{-(t-2T)/\tau} \right\}, \quad \text{for } 2T < t < 4T$$

and the load voltage

$$v_L(t) = 0, \text{ for } t < T$$

$$v_L(t) = V_G(1 - e^{-(t-T)/\tau}), \text{ for } T < t < 3T$$

$$v_L(t) = V_G \left\{ 1 - e^{-(t-T)/\tau} + \frac{t - 3T}{\tau} e^{-(t-3T)/\tau} \right\}, \text{ for } 3T < t < 5T$$

for the first two time intervals of interest.

For illustration, we present results using specific numerical values for the components. Suppose that $V_G = 5$ V, $T = 100$ ps, $Z_0 = 50$ Ω, and $L = 2$ nH. For these values, the time constant associated with the response of the system is

$$\tau = L/(2Z_0) = 2 \times 10^{-11} \text{ s} = 20 \text{ ps} \qquad (8.44)$$

which is one fifth of the transit delay T. Figure 8.9 shows a plot of the input voltage and current for these values. Under these conditions, the inductance contributes an appreciable delay to the signal launched on the line.

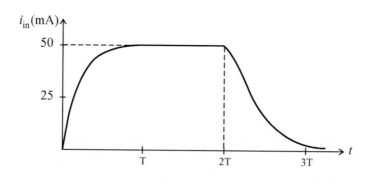

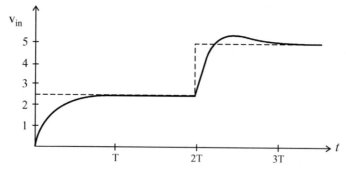

FIGURE 8.9: Plot of input current and voltage.

From the circuit at the generator end of the line in Figure 8.8, or equivalently the relation $V_{\text{chip}} = v_{\text{in}} + Z_0 I_{\text{in}}$, we may also determine the value of the supply voltage seen at the local (chip) level of the circuit. For the first two time intervals, this voltage is given by

$$V_{\text{chip}}(t) = V_G(1 - e^{-t/\tau}), \quad \text{for } 0 < t < 2T \tag{8.45}$$

$$V_{\text{chip}}(t) = \frac{V_G}{2}\left\{1 - e^{-t/\tau} + \frac{t - 2T}{\tau}e^{-(t-2T)/\tau}\right\}, \quad \text{for } 2T < t < 4T \tag{8.46}$$

A plot of V_{chip} is shown in Figure 8.10. As the driver initially draws current from the supply, the local supply voltage drops due to the inductance associated with the path from the supply to the driver. Physically, this is related to the fact that there is insufficient charge stored on the chip to supply the current drawn by the line. Since V_{chip} is the actual supply voltage experienced not only by the driver gate in question but by other gates in the immediate vicinity, voltage fluctuations of this sort perturb signals on other parts of the chip as well, raising the noise level of the entire system. When large numbers of gates switch in unison, the resulting simultaneous switching noise can be substantial. We note that additional capacitance, in the form of decoupling capacitors, can be added in the immediate vicinity of the driver to correct this problem.

In this chapter, we considered several examples involving transmission lines with reactive loads. Reactive parasitics can prevent the proper operation of high-speed digital systems, and their analysis is necessary to enable compensation during the design stage. For instance, decoupling capacitors can be added to the power supply network in order to compensate for the equivalent inductance illustrated above.

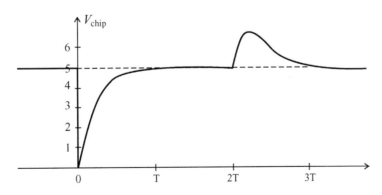

FIGURE 8.10: Plot of V_{chip}.

REFERENCES

[1]   S. Rosenstark, *Transmission Lines in Computer Engineering*. New York: McGraw-Hill, 1994.

[2]   J. A. Strickland, *Time–Domain Reflectometry Measurements*. Beaverton, OR: Tektronix, 1970.

PROBLEMS

8.1   *Fill in the blank:* In an uncharged network with a switched DC excitation, capacitors initially look like _____ circuits, and inductors initially look like _____ circuits.

8.2   *Fill in the blank:* In a network with a switched DC excitation, a long time after the switch has been closed and the voltages and currents reach steady state, capacitors look like _____ circuits, and inductors look like _____ circuits. .

8.3   The transmission line and capacitor in the system shown below are initially uncharged. The switch in the system is closed at $t = 0$ and remains closed thereafter.

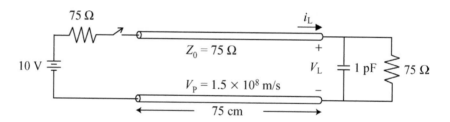

(a)   Determine the one-way transit time T for the line.

(b)   Formulate and solve a differential equation to obtain an expression for the load voltage $V_L(t)$ during the time $t > T$.

(c)   At what instant in time does the load voltage exceed 4.0 V? If the load end of this line represented a gate in a digital system, and that gate triggered at 4.0 V, how much additional delay is due to the capacitor charge time?

(d)   Determine an expression for the load current $I_L(t)$ for $t > T$

8.4   The system used in Problem 8.3 has been operated for a long time with the switch in the closed position, and the transmission line and the load capacitor are both charged to 5.0 V. Now, suppose that the switch is opened at time $t = t_0$. Formulate and solve a differential equation to obtain an expression for the load voltage $V_L(t)$ for time $t > t_0 + T$.

8.5   The transmission line system depicted below represents the connection between two logic gates and includes parasitic capacitance at the receiver. Switch A in the system is closed for a period of 30 ns to represent a short logic pulse being sent down the line.

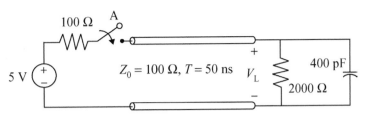

(a)   Sketch the load voltage $V_L(t)$ during the time $t > 0$, clearly labeling all voltage levels and times. You may ignore reflections from the source end of the line after the switch is opened.

(b)   If the logic pulse will not trigger the receiver gate until the voltage exceeds 2.5 V, what is the total triggering latency (as measured from the initial switch closing at $t = 0$) for this system?

8.6   The transmission line and capacitor in the system shown below are initially uncharged. The generator voltage consists of the finite-duration square pulse shown in the figure. Determine an expression for the load current $I_L(t)$ for $t > T$. Plot the load current.

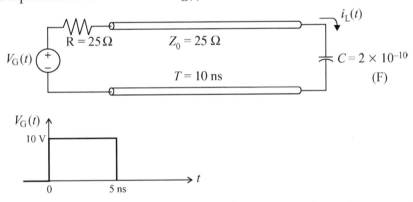

8.7   Henry is designing a clock circuit that outputs a short square pulse to drive several microprocessors. When Henry finishes etching his printed circuit board and soldering chips onto the board, he finds that the output of his clock, as measured from the input terminals of his application-specific integrated circuit chips, not a clean square pulse; instead, the output looks like the graph below:

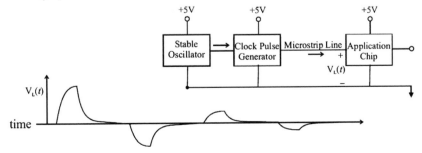

Using what you know so far about transmission lines, diagnose two distinct problems with this circuit.

8.8   The transmission line and inductor in the system shown below are initially unenergized. The switch in the system is closed at $t = 0$ and remains closed thereafter.

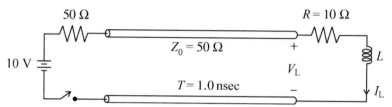

Formulate and solve a differential equation to obtain an expression for the load current $I_L(t)$ during the time $t > T$. Plot $I_L(t)$ for $0 < t < 3T$.

8.9   The system shown below is switched through a sequence of states:

State 0: Both switches are open, and both lines are uncharged.

State 1: Immediately after switch A is closed.

State 2: Switch A has been closed for a long time.

State 3: Immediately after switch B is closed.

State 4: Switch B has been closed for a long time.

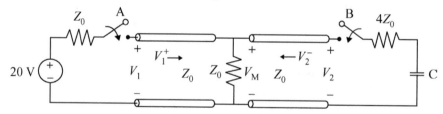

Fill out the following table according to these switching states. Assume that all backward-propagating waves are measured from the rightmost side of the transmission line. Assume all forward-propagating waves are measured from the leftmost side of the transmission line.

	V_1	V_M	V_2	V_1^+	V_2^-
State 0	0	0	0	0	0
State 1		0	0		0
State 2					
State 3					
State 4					

8.10 The system shown below is switched through a sequence of states:

State 0: Both switches are open and both lines are uncharged.

State 1: Immediately after switch A is closed.

State 2: Switch A has been closed for a long time.

State 3: Immediately after switch B is closed.

State 4: Switch B has been closed for a while.

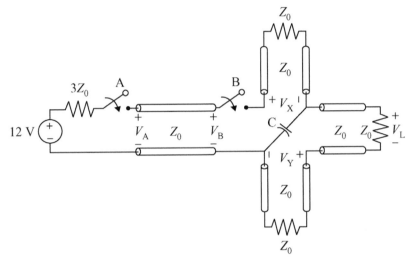

Fill in the following table according to these switching states:

	V_A	V_B	V_L	V_X	V_Y
State 0	0	0	0	0	0
State 1					
State 2					
State 3					
State 4					

8.11 The system shown below is switched through a sequence of states:

State 0: All switches are open and all lines are uncharged.

State 1: Immediately after switch A is closed.

State 2: Switch A has been closed for a long time.

State 3: Immediately after switch B is closed.

State 4: Switch B has been closed for a long time.

State 5: Immediately after switch C is closed.

State 6: Switch C has been closed for a long time.

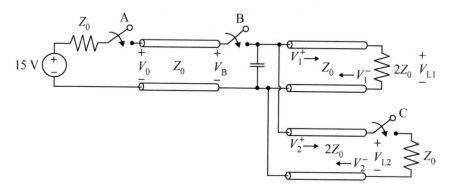

Fill in the following table according to these switching states. Note that all positive-going waves are measured at the generator side and all negative-going waves are measured at the load side.

	V_0	V_1^+	V_1^-	V_2^+	V_2^-	V_{L1}	V_{L2}	V_B
State 0	0	0	0	0	0	0	0	0
State 1								
State 2								
State 3								
State 4								
State 5								
State 6								

CHAPTER 9

Lines with Nonlinear Loads

Objectives: Extend transmission line analysis to lines having nonlinear terminations. Consider both the iterative numerical solution of nonlinear problems and the graphical solution. Illustrate for an example with logic gates described by a graphical V–I curve.

A resistively terminated transmission line system is shown in Figure 9.1. In previous chapters, we approached a resistively loaded system by solving the appropriate equations analytically, introducing reflection coefficients and other simplifications. In the event that the load is a nonlinear device, however, we must revert back to first principles in order to work out the solution. For a resistive load, a linear device, the appropriate equations describing the load voltage and current are given by

$$I_L = \frac{V_L}{R_L} \tag{9.1}$$

and

$$I_L = \frac{2V^+}{Z_0} - \frac{V_L}{Z_0} \tag{9.2}$$

Equation (9.2) can be derived from Equations (3.9) and (3.10) by eliminating V^-. Equation (9.1) describes the voltage/current relationship at the load, while Equation (9.2) describes the constraint imposed on the current and voltage by the transmission line.

Other than an analytical solution, there are two approaches we might attempt to solve Equations (9.1) and (9.2). The first idea is to substitute one equation into the other, yielding either

$$I_L = \frac{2V^+}{Z_0} - \frac{I_L R_L}{Z_0} \tag{9.3}$$

or

$$\frac{V_L}{R_L} = \frac{2V^+}{Z_0} - \frac{V_L}{Z_0} \tag{9.4}$$

Obviously, either of these equations could be solved analytically for I_L or V_L. However, consider instead an iterative numerical solution, such as might be accomplished with the aid of a calculator

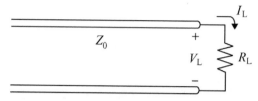

FIGURE 9.1: A resistively terminated transmission line.

or personal computer. A "guess" for I_L can be substituted into the right-hand side of Equation (9.3), after which a new estimate of I_L is obtained by evaluating the right-hand expression. In other words, we use the update equation

$$I_L^{new} = \frac{2V^+}{Z_0} - \frac{I_L^{old} R_L}{Z_0} \qquad (9.5)$$

A similar iterative equation can be based on Equation (9.4) and solved for V_L. Or, these equations can be rearranged so that quantities on the left-hand side represent the "old" values, as might sometimes be necessary to ensure convergence of the iterative procedure.

The second approach to solving Equations (9.1) and (9.2) is to construct a solution graphically. Figure 9.2 illustrates a plot of these two equations, on an I_L versus V_L scale. In this linear situation, Equation (9.1) yields a straight line with a slope of $(1/R_L)$, while Equation (9.2) is also a straight line with a slope of $(-1/Z_0)$. The intersection point is the solution.

Figure 9.3 shows a transmission line terminated in a nonlinear device, having $V–I$ characteristic

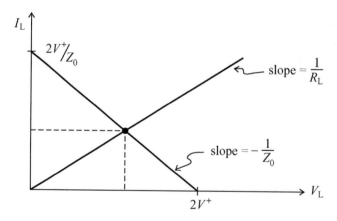

FIGURE 9.2: A plot of Equations (9.1) and (9.2) to facilitate a graphical solution.

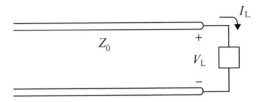

FIGURE 9.3: A transmission line terminated in a nonlinear load.

$$V_L = 50(I_L)^2, \ I_L > 0 \qquad (9.6)$$

Suppose for specificity the line has $Z_0 = 50 \ \Omega$ and $V^+ = 5$ V. Equations (9.6) and (9.2) can be combined to produce

$$
\begin{aligned}
V_L^{\text{new}} &= 50 \left(\frac{2V^+}{Z_0} - \frac{V_L^{\text{old}}}{Z_0} \right)^2 \\
&= \frac{\left(10 - V_L^{\text{old}} \right)^2}{50}
\end{aligned}
\qquad (9.7)
$$

Suppose also that we estimate the solution as $V_L = 2.5$ V. An iterative process can be carried out using Equation (9.7), yielding (in volts)

initial estimate: $V_L = 2.5$

iteration 1: $V_L = 1.125$

iteration 2: $V_L = 1.575$

iteration 3: $V_L = 1.420$

iteration 4: $V_L = 1.472$

iteration 5: $V_L = 1.454$

iteration 6: $V_L = 1.461$

iteration 7: $V_L = 1.458$

iteration 8: $V_L = 1.459$

iteration 9: $V_L = 1.459$

After eight iterations, the result converged to four decimal places. (This is somewhat lucky—it does not always work so well!) The corresponding load current is found from Equation (9.6) to be $I_L = 0.171$ A. Given these values, we can easily calculate the reflected voltage

$$V^- = V_L - V^+ = -3.541 \text{ V} \qquad (9.8)$$

and the effective reflection coefficient for this specific value of V^+

$$\Gamma_L = V^- / V^+ = -0.7082 \qquad (9.9)$$

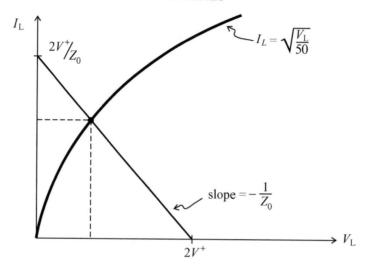

FIGURE 9.4: Graphical solution of Equations (9.2) and (9.6).

For this V^+, the load is equivalent to an 8.53-Ω resistor.

Alternatively, we could seek a graphical solution. Figure 9.4 shows a plot of Equations (9.2) and (9.6) on an I_L versus V_L scale. The solution can be read from the intersection point on the graph. The accuracy of the solution obtained graphically depends, of course, on the resolution of the graph.

To analyze a transmission line system, the entire solution process may require additional steps to account for reflections from the ends of the line. Figure 9.5 shows a complete transmission line system, terminated with the nonlinear load illustrated above. If the switch in this system is closed at $t = 0$, the load voltage will jump from 0 to 1.459 V at time $t = T$, as determined above. This load voltage remains constant during the interval $T < t < 3T$. The reflected voltage ($V^- = -3.541$) produced during this time interval propagates back to the generator end of the line, where it encounters a reflection coefficient of $\Gamma_G = -1$. The additional reflection from the generator end of the line during

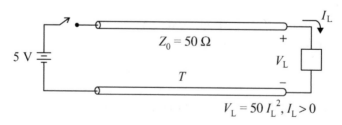

FIGURE 9.5: A transmission line system with generator, switch, and nonlinear load.

this interval must be added with the 5-V signal produced by the battery. Thus, at $t = 3T$, the value of V^+ at the load changes to 8.541 V, meaning that the equation to be solved for the load voltage also changes, to

$$V_L^{new} = \frac{\left(17.082 - V_L^{old}\right)^2}{50} \tag{9.10}$$

We note that it is necessary to work with the *total* incident voltage, since the load is nonlinear and we are no longer able to use the concept of superposition when solving for the load voltage.

Using an initial estimate of $V_L = 3.5$ V, we iterate using Equation (9.10) to produce (in volts)

iteration 1: $V_L = 3.689$

iteration 2: $V_L = 3.587$

iteration 3: $V_L = 3.642$

•

•

•

iteration 9: $V_L = 3.623$

iteration 10: $V_L = 3.623$

Thus, after 10 iterations, we determine the load voltage, the load current $I_L = 0.269$ A, and the new total reflected voltage $V^- = -4.918$ V for the time interval $3T < t < 5T$. These values also dictate the total incident voltage during the subsequent time interval $5T < t < 7T$, given by $V^+ = 9.918$.

During the interval $5T < t < 7T$, the equation to be solved for the load voltage is

$$V_L^{new} = \frac{\left(19.836 - V_L^{old}\right)^2}{50} \tag{9.11}$$

Using a procedure essentially identical to the above, based on the initial estimate $V_L = 4.5$ V, we generate the following (in volts):

iteration 1: $V_L = 4.704$

iteration 2: $V_L = 4.580$

iteration 3: $V_L = 4.655$

•

•

•

iteration 15: $V_L = 4.627$

iteration 16: $V_L = 4.627$

Furthermore, we determine the load current during the interval $5T < t < 7T$ as $I_L = 0.304$ A.

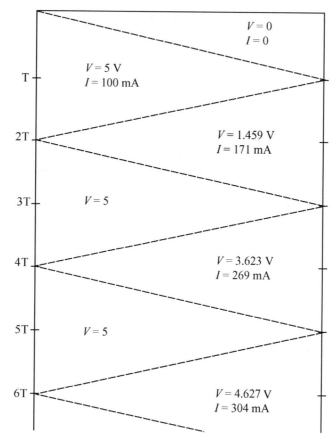

$V = 0$
$I = 0$

T

$V = 5$ V
$I = 100$ mA

2T

$V = 1.459$ V
$I = 171$ mA

3T

$V = 5$

4T

$V = 3.623$ V
$I = 269$ mA

5T

$V = 5$

6T

$V = 4.627$ V
$I = 304$ mA

FIGURE 9.6: A reflection diagram showing the total voltages and currents on the line of Figure 9.5 for $0 < t < 6T$.

The iterative process can be continued for as many time intervals as desired. In this example, the load voltage will eventually converge to $V_L = 5$ V, and the load current will converge to $I_L = 0.316$ A. For convenience, the results can be collected on a reflection diagram, as illustrated in Figure 9.6.

As a second and perhaps more practical example, we turn our attention to a graphical solution technique. Figure 9.7 depicts a transmission line system consisting of two gates connected by a line with characteristic impedance $Z_0 = 50$ Ω. Figure 9.8 depicts the I–V characteristics of the gates in question, TTL 7404 inverters. At time $t = 0$, the initial state of the line is logic zero.

From Figure 9.8, it is apparent that the logic zero state occurs when the line is charged to a voltage of about 0.1 V and a line current of about −1.0 mA (the negative current is typical of

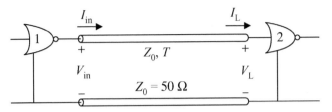

FIGURE 9.7: Two logic gates connected by a transmission line.

TTL logic). Thus, initially $V_{in} = V_L \cong 0.1$ and $I_{in} = I_L \cong -0.001$. By decomposing these values into forward- and reverse-going waves (Chapter 6), we find that $V^+ = 0.025$ and $V^- = 0.075$.

At time $t = 0$, the state of the driver changes from logic zero to logic 1, and we must determine the new forward-going voltage. Figure 9.9 shows an equivalent circuit for the generator end of the line at time $t = 0^+$. The Thevenin equivalent circuit is equivalent to a V–I characteristic of

$$V_{in} = 50 I_{in} + 0.15 \qquad (9.12)$$

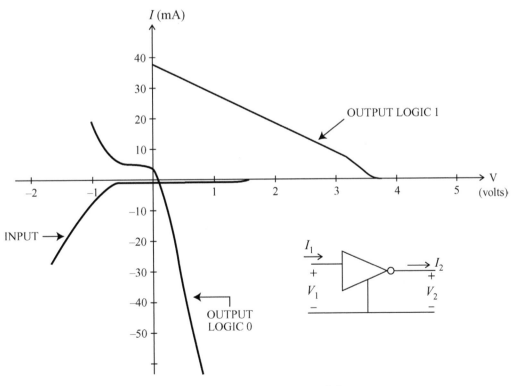

FIGURE 9.8: V–I characteristics for a TTL 7404 inverter gate [1].

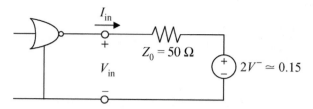

FIGURE 9.9: Equivalent circuit for the generator end of the line at $t = 0^+$.

Figure 9.10 shows an I versus V plot of the two appropriate characteristic curves representing the generator end of the line at this time. One of these curves is the "logic one" output characteristic for the device; the other is Equation (9.12). The intersection of these two curves occurs at $V_{in} \cong 1.35$ V and $I_{in} \cong 25$ mA. Consequently, at $t = 0$ V_{in} changes from 0.1 to 1.35 V. Furthermore, the positive-going waveform on the line jumps from a steady DC voltage of 0.025 to 1.425 V, producing a new value ($V^+ = 1.425$) at the load end of the line beginning at time $t = T$. (We should probably round this result to 1.4 V given the resolution of the graph!)

An equivalent circuit for the load end of the line at time $t = T^+$ is depicted in Figure 9.11. Figure 9.12 shows a graph of the two relevant curves, the input characteristic for the device and the circuit equation from the equivalent circuit of Figure 9.11,

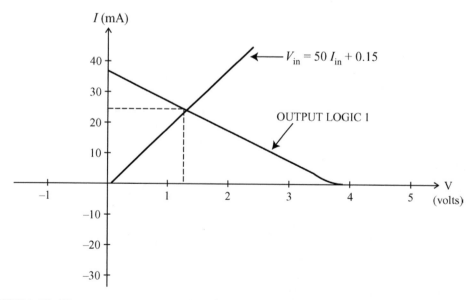

FIGURE 9.10: The two curves representing the V–I characteristics of the generator end of the line at time $t = 0^+$.

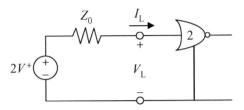

FIGURE 9.11: Equivalent circuit for the load end of the line at $t = T^+$.

$$I_L = \frac{2V^+}{Z_0} - \frac{V_L}{Z_0}$$
$$= 0.056 - \frac{V_L}{50} \tag{9.13}$$

The intersection of these curves on the graph yields $V_L \cong 2.8$ V and $I_L \cong 0.0$ mA. (Observe that the gate appears to be an open circuit across the line for voltages in excess of 1.5 V.) Therefore, beginning at time $t = T$, there is a reflection generated at the load end of the line with voltage $V^- \cong 1.4$ V and $I^- \cong -28$ mA.

At time $t = 2T$, the reflected voltage of $V^- \cong 1.4$ V arrives back at the generator end of the line, producing a new equivalent circuit as depicted in Figure 9.13. The graphical situation is plotted in Figure 9.14 and leads to a solution of $V_{in} = 3.2$ V and $I_{in} = 7.5$ mA for the time interval $2T < t < 4T$. The positive going waveform also changes at $t = 2T$, with a new V^+ equal to 1.8 V. This new V^+ arrives at the load end of the line at $t = 3T$.

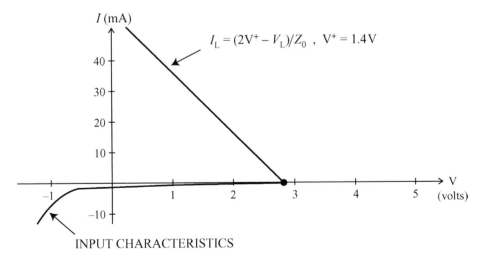

INPUT CHARACTERISTICS

FIGURE 9.12: V–I characteristics for load end of the line at $t = T$.

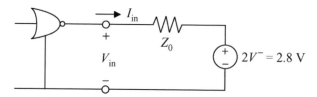

FIGURE 9.13: Equivalent circuit for generator end, $t = 2T$.

At time $t = 3T$, the total positive-going voltage at the load end of the line changes to $V^+ \cong$ 1.8 V. Since the load appears to be an open circuit (see Figure 9.12), we obtain $V_L \cong 3.6$ V and $I_L \cong$ 0.0 mA for the time interval $3T < t < 5T$. A new total reflected voltage of $V^- \cong 1.8$ V is generated beginning at $t = 3T$.

At $t = 4T$, the reflected voltage $V^- \cong 1.8$ V arrives at the generator end of the system, shifting the plot of the equation for the input circuit to the right on the graph (Figure 9.15). The new result obtained for the time interval $4T < t < 6T$ is $V_{in} \cong 3.6$ V and $I_{in} \cong 0$ mA.

In this example, the line voltage will stabilize somewhere above 3.6 V. Our resolution is limited to some extent by the available plot for the device characteristic curves, so we are unable to push the analysis beyond this point. The transient voltages and currents can be arranged on a reflection diagram for convenience, as illustrated in Figure 9.16. As in previous lectures, the reflection diagram provides the history of the transient process and can be used for generating plots of the load voltage as a function of time, etc. We observe that the device mismatch in this example delays the second gate triggering time until $t = 3T$.

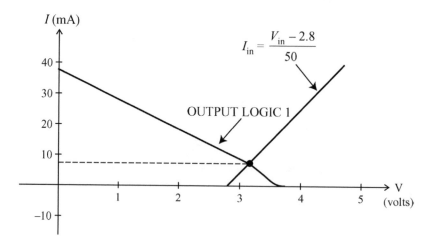

FIGURE 9.14: V–I characteristics for generator end at time $t = 2T$.

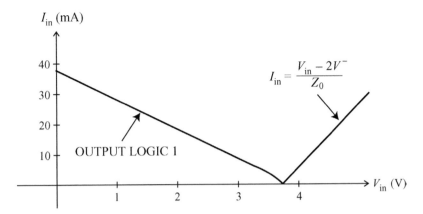

FIGURE 9.15: *V–I* curves for generator end, $t = 4T$.

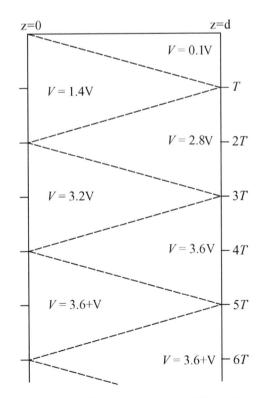

FIGURE 9.16: A reflection diagram for the two-gate system. The voltage stabilizes somewhere above 3.6 V, where the curves in Figure 9.8 cross.

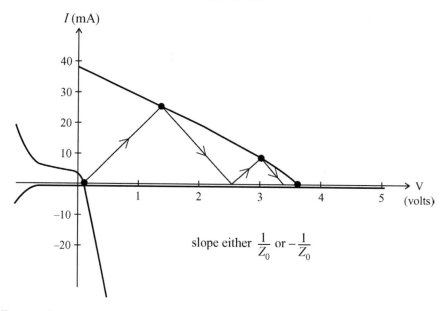

FIGURE 9.17: Composite graph showing solutions for the successive load and generator voltages and currents, obtained by drawing lines with slopes of either $+1/Z_0$ or $-1/Z_0$ to move from one result to the next.

The process of developing the transient solution for this nonlinear example involves the repeated graphical solution of the intersection points of (1) the characteristic curves for the device and (2) the straight lines representing the transmission-line equivalent circuits at the generator and load ends. For each time interval, these lines have the same slope but different intercepts. By combining the graphs from Figures 9.10, 9.12, 9.14, and 9.15, we arrive at a composite graph shown in Figure 9.17. Figure 9.17 illustrates that the graphical solution may be obtained by drawing lines with slopes alternating between $+1/Z_0$ and $-1/Z_0$, connecting solutions from previous time intervals to successive results. Once the reader develops a facility for the graphical solution, the "shortcut" procedure suggested in Figure 9.17 may be employed to obtain the graphical solution from a single plot of the device characteristics.

In this chapter, we have considered the treatment of nonlinear loads, using two procedures that are generalizations of the analytical solution methods employed in the linear situation. In practice, a wide variety of nonlinear terminations can be treated in this manner. The results from this analysis can be arranged on a reflection diagram and interpreted in a manner similar to results obtained for linear devices or terminations.

REFERENCE

[1] S. P. Castillo, *Electromagnetic Modeling of High Speed Digital Circuits*. PhD dissertation, University of Illinois, Urbana, IL, 1987.

PROBLEMS

9.1 *Fill in the blank:* An _____ numerical procedure can be used for solving for voltages and currents in a circuit with nonlinear components.

9.2 *Fill in the blank:* We cannot use superposition to calculate individual reflections on a transmission line with a diode load, because the diode is a _____ device.

9.3 *True or false:* The V–I characteristic of a resistor is a straight line.

9.4 The lossless transmission line in the figure below is terminated with the nonlinear load described by the given V–I characteristic. The transmission line is initially uncharged and the switch is closed at time $t = 0$.

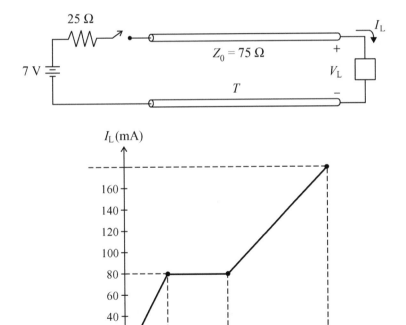

(a) Complete the reflection diagram to find voltages V_a, V_b, V_c, and V_d:

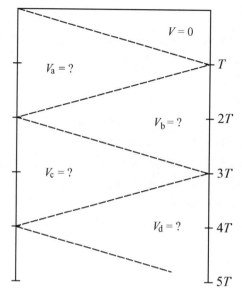

(b) Plot the load voltage and current for $0 < t < 5T$, labeling all amplitudes.

9.5 The lossless transmission line in the figure below is terminated with the nonlinear load described by the given V–I characteristic. The transmission line is initially uncharged and the switch is closed at time $t = 0$.

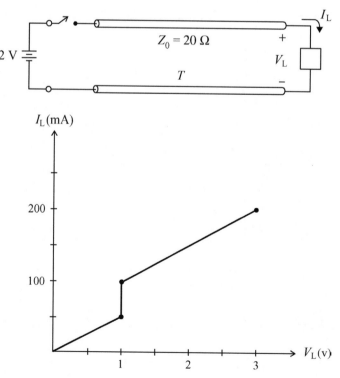

Plot the load voltage and current for $0 < t < 5T$, labeling all amplitudes.

9.6 The transmission line system shown below is terminated with a diode whose $V\!-\!I$ characteristic is described by the equation

$$V_L = V_0 \ln\left(\frac{I_L}{I_0} + 1\right)$$

where $V_0 = 0.1$ V and $I_0 = 1$ mA.

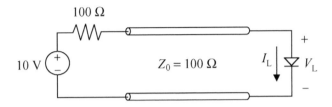

(a) Find the steady-state load voltage and current, using the iterative computational procedure.

(b) Calculate the steady-state reflected voltage on the line.

(c) Now, reverse the diode at the load end of the line, and find the new steady state load voltage and current. (Note that the $V\!-\!I$ characteristic equation must be modified for the reversed diode.)

(d) Calculate the steady-state reflected voltage on the line for the reversed diode.

9.7 It is known that for certain loads and excitations, nonlinear systems (and circuits) can often become *chaotic*. Transmission lines are no different: if you connect a source and certain types of nonlinear loads, the load voltage may never reach a steady state; instead, it exhibits chaotic or pseudostochastic behavior. Find an example of a transmission line with a nonlinear load and provide a graph of its chaotic output, $V_L(t)$. Cite any book pages or journal articles you consult.

CHAPTER 10

Crosstalk on Weakly Coupled Transmission Lines

Objectives: Develop the equations describing coupling between identical, lossless transmission lines in a homogeneous environment. Assuming that the coupling is weak, find the general solution for the signal coupled into line 2 by an arbitrary signal on line 1. Illustrate with several examples.

A common source of noise in both analog and digital systems arises from parasitic mutual capacitance and inductance between closely spaced conductors. Mutual capacitance and inductance provides a path between two traces that are not physically connected, allowing signals to jump from one circuit to another. In the following, we develop a methodology for analyzing crosstalk in coupled transmission lines.

Figure 10.1 shows the cross section of two closely spaced traces in a uniform, homogeneous environment. In common with the analysis of an individual transmission line, the analysis of coupled lines is facilitated through an equivalent circuit for an infinitesimally short length of coupled lines. Such a circuit is depicted in Figure 10.2, under the assumption that the lines are identical and lossless. This circuit includes elements representing mutual capacitance and inductance.

An application of Kirchhoff's laws to the equivalent circuit in Figure 10.2 produces the following four coupled equations, involving the voltages $v_i(t,z)$ and currents $i_i(t,z)$.

$$\frac{\partial v_1}{\partial z} = -L_s \frac{\partial i_1}{\partial t} - L_m \frac{\partial i_2}{\partial t} \qquad (10.1)$$

$$\frac{\partial i_1}{\partial z} = -C_s \frac{\partial v_1}{\partial t} - C_m \frac{\partial v_1}{\partial t} + C_m \frac{\partial v_2}{\partial t} \qquad (10.2)$$

$$\frac{\partial v_2}{\partial z} = -L_s \frac{\partial i_2}{\partial t} - L_m \frac{\partial i_1}{\partial t} \qquad (10.3)$$

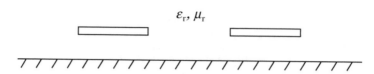

FIGURE 10.1: A cross section of two identical traces over a ground plane.

$$\frac{\partial i_2}{\partial z} = -C_s \frac{\partial v_2}{\partial t} - C_m \frac{\partial v_2}{\partial t} + C_m \frac{\partial v_1}{\partial t} \tag{10.4}$$

These equations can be rewritten in terms of four parameters given by the total capacitance per unit length

$$C = C_s + C_m \tag{10.5}$$

the mutual per unit length capacitance

$$E = C_m \tag{10.6}$$

the per unit length self inductance

$$L = L_s \tag{10.7}$$

and the mutual per unit length inductance

$$M = L_m \tag{10.8}$$

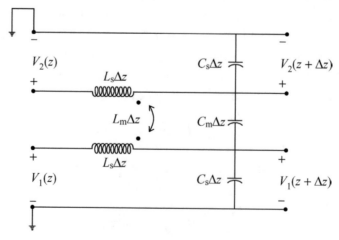

FIGURE 10.2: A short length of coupled, lossless transmission lines.

The resulting equations are

$$\frac{\partial v_1}{\partial z} = -L\frac{\partial i_1}{\partial t} - M\frac{\partial i_2}{\partial t} \tag{10.9}$$

$$\frac{\partial i_1}{\partial z} = -C\frac{\partial v_1}{\partial t} + E\frac{\partial v_2}{\partial t} \tag{10.10}$$

$$\frac{\partial v_2}{\partial z} = -L\frac{\partial i_2}{\partial t} - M\frac{\partial i_1}{\partial t} \tag{10.11}$$

$$\frac{\partial i_2}{\partial z} = -C\frac{\partial v_2}{\partial t} + E\frac{\partial v_1}{\partial t} \tag{10.12}$$

Under the assumption that the medium surrounding the traces is uniform or homogeneous, physical constraints (stemming from the fact that electrical signals must propagate at the velocity of light) dictate that

$$\frac{M}{L} = \frac{E}{C} \tag{10.13}$$

We will refer to this unitless ratio as Δ, in other words,

$$\Delta = \frac{M}{L} = \frac{E}{C} \tag{10.14}$$

A second constraint imposed by the velocity of light in a uniform medium is

$$LC - ME = LC\left(1 - \Delta^2\right) = \mu\varepsilon \tag{10.15}$$

where μ and ε denote the total permeability and permittivity of the medium, respectively. For a microstrip structure, a similar constraint can be expressed in terms of the effective permittivity. For a given coupled transmission line geometry, the four parameters L, M, C, and E can be determined by analysis or measurement.

We are primarily interested in the situation depicted in Figure 10.3, where line 1 is excited by a generator with voltage V_G and line 2 is nearby but not directly excited. When discussing coupled lines, line 1 is often referred to as the *aggressor* line, while line 2 is the *victim* line. In the case of strong coupling between the two lines, the presence of line 2 will distort the signal on line 1, complicating the analysis. We will assume that the coupling is weak, meaning that

$$\Delta = \frac{M}{L} = \frac{E}{C} \ll 1 \tag{10.16}$$

This is a very good assumption for most transmission lines; in fact, it is difficult to obtain strong coupling without deliberately designing for it. (There are devices such as directional couplers and multiplexers that rely on strong coupling.)

Under the assumption in Equation (10.16), the presence of the second line has a negligible effect on the waveform on line 1. Therefore, line 1 can be analyzed using ordinary transmission line theory as previously developed for a single line, using the self-inductance L and self-capacitance C to describe the line parameters. In other words, Equations (10.9) and (10.10) are approximated by

$$\frac{\partial v_1}{\partial z} \cong -L\frac{\partial i_1}{\partial t} \qquad (10.17)$$

$$\frac{\partial i_1}{\partial z} \cong -C\frac{\partial v_1}{\partial t} \qquad (10.18)$$

Observe that the lines in Figure 10.3 are terminated with loads equal to their characteristic imped-ance, eliminating reflections.

Even in the weakly coupled case, a waveform on line 1 will induce a signal into line 2, and our remaining objective is to determine this induced signal. We will approach this problem in the same way as Rosenstark [1], by considering the coupling to a short segment of line 2, having length Δz, and using the portions of Equations (10.11) and (10.12) given by

$$\frac{\partial v_2}{\partial z} \cong \frac{\Delta v_2}{\Delta z} \cong -M\frac{\partial i_1}{\partial t} \qquad (10.19)$$

$$\frac{\partial i_2}{\partial z} \cong \frac{\Delta i_2}{\Delta z} \cong E\frac{\partial v_1}{\partial t} \qquad (10.20)$$

Replacing Equations (10.11) and (10.12) by these approximations might appear to be questionable, since the discarded terms seemingly are of the same order of magnitude as those retained. After

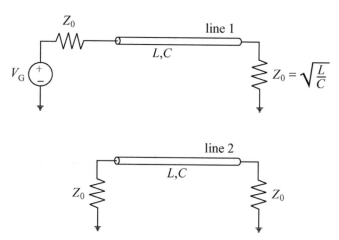

FIGURE 10.3: Two weakly coupled transmission lines.

we develop a solution to these equations, however, we will justify their validity by showing that the solution satisfies Equations (10.11) and (10.12) to the expected order of the approximation.

We will separately consider capacitive and inductive coupling between the lines, using the equivalent circuit of Figure 10.2. Capacitive coupling injects a current i_C into the segment of line 2 under consideration, as described by Equation (10.20) and depicted in Figure 10.4. From Figure 10.4, we deduce that $\Delta i_2 = i_2(z + \Delta z) - i_2(z) = i_c$, and from (10.20) write

$$i_C = E \frac{\partial v_1}{\partial t} \Delta z \tag{10.21}$$

This additional current excites forward- and reverse-going waves in line 2, produced at the location of the short segment under consideration. These waves propagate away from that location with incremental values

$$\Delta v_2^+ = \frac{i_C Z_0}{2} = \frac{E \Delta z Z_0}{2} \frac{\partial v_1}{\partial t} \tag{10.22}$$

$$\Delta v_2^- = \frac{i_C Z_0}{2} = \frac{E \Delta z Z_0}{2} \frac{\partial v_1}{\partial t} \tag{10.23}$$

The inductive coupling situation involves an additional series voltage induced in line 2, as illustrated in Figure 10.5. Using Equation (10.19), we infer that $\Delta v_2 = v_L$, and therefore,

$$v_L = -M \frac{\partial i_1}{\partial t} \Delta z \tag{10.24}$$

Since it is more convenient to express quantities in terms of the voltage on line 1 than the current, we make the additional assumption at this point that the waveform on line 1 represents a forward-going signal, and therefore, Equation (10.24) can be rewritten as

$$v_L = -M \frac{1}{Z_0} \frac{\partial v_1}{\partial t} \Delta z \tag{10.25}$$

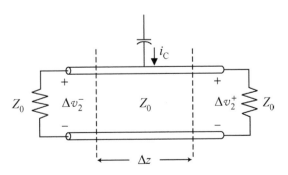

FIGURE 10.4: Illustration of capacitive coupling into line 2.

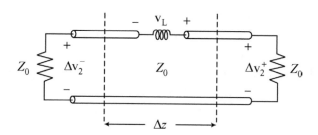

FIGURE 10.5: Illustration of inductive coupling into line 2.

From Figure 10.5, we conclude that this excitation creates a forward-going wave with incremental voltage

$$\Delta v_2^+ = -\frac{M \Delta z}{2 Z_0} \frac{\partial v_1}{\partial t} \qquad (10.26)$$

and a reverse-going wave with incremental voltage

$$\Delta v_2^- = \frac{M \Delta z}{2 Z_0} \frac{\partial v_1}{\partial t} \qquad (10.27)$$

By combining the capacitive and inductive contributions, we obtain

$$\Delta v_2^+ = \left(E Z_0 - \frac{M}{Z_0} \right) \frac{\Delta z}{2} \frac{\partial v_1}{\partial t} \qquad (10.28)$$

$$\Delta v_2^- = \left(E Z_0 + \frac{M}{Z_0} \right) \frac{\Delta z}{2} \frac{\partial v_1}{\partial t} \qquad (10.29)$$

These are the incremental contributions produced at the segment of line 2 under consideration by a signal at the same location and time on line 1. For convenience, we rewrite these expressions in the form

$$\Delta v_2^+ = \kappa_F \Delta z \frac{\partial v_1}{\partial t} \qquad (10.30)$$

$$\Delta v_2^- = \kappa_R \frac{2}{v_p} \Delta z \frac{\partial v_1}{\partial t} \qquad (10.31)$$

where, following the convention of Rosenstark [1], we have introduced forward and reverse coupling coefficients

$$\kappa_F = \frac{E Z_0}{2} - \frac{M}{2 Z_0} = \frac{\sqrt{LC}}{2} \left(\frac{E}{C} - \frac{M}{L} \right) \qquad (10.32)$$

$$\kappa_R = \frac{1}{2\sqrt{LC}} \left(\frac{E Z_0}{2} + \frac{M}{2 Z_0} \right) = \frac{1}{4} \left(\frac{E}{C} + \frac{M}{L} \right) \qquad (10.33)$$

and where $v_p = 1/\sqrt{LC}$. The parameter κ_F has units of time per distance, while κ_R is unitless. If the lines are symmetric, the medium is homogeneous, and the coupling is weak, it follows from our earlier assumptions that

$$\frac{E}{C} - \frac{M}{L} = 0 \tag{10.34}$$

and

$$\frac{E}{C} + \frac{M}{L} = 2\Delta \tag{10.35}$$

Consequently, $\kappa_F = 0$ and $\kappa_R = \Delta/2$. However, in the more general situation of nonsymmetric lines, a non-uniform medium, etc., the coefficient κ_F will not vanish. Therefore, for generality we will retain both coefficients throughout the following development.

The results in Equations (10.30) and (10.31) represent the coupling to a short segment of line 2 induced by a forward-going signal at the same location (space and time) along line 1. To obtain the complete coupling to line 2, in order to determine the terminal voltages, we must superimpose the effects of coupling along the entire line.

In general, the waveform on line 1 has the form

$$v_1(t, z) = V^{\text{inc}}\left(t - \frac{z}{v_p}\right) \tag{10.36}$$

The forward crosstalk produced by this source signal at location z' on line 1 is subsequently observed at a location $z > z'$ on line 2. At that location, which involves the additional propagation delay $(z - z')/v_p$, the argument of Equation (10.30) is modified to

$$t - \frac{z'}{v_p} - \frac{(z - z')}{v_p} = t - \frac{z}{v_p} \tag{10.37}$$

Therefore, the contribution to the forward crosstalk at location z is given by

$$\Delta v_2^+ = \kappa_F \Delta z \left. \frac{\partial V^{\text{inc}}}{\partial t} \right|_{t - \frac{z}{v_p}} \tag{10.38}$$

Equation (10.38) includes just the coupling at the short segment. By superimposing the effects of forward coupling along the entire line (between the generator end and location z), we obtain the signal at location z as

$$v_2(z, t) = \kappa_F \int_{z'=0}^{z} \frac{\partial V^{\text{inc}}}{\partial t} dz'$$

$$= \kappa_F \left(\left. \frac{\partial V^{\text{inc}}}{\partial t} \right|_{t - \frac{z}{v_p}} \right) z \tag{10.39}$$

For reverse crosstalk, we again consider an incident wave on line 1 with the form of (10.36). The reverse-propagating signal induced at location z' on line 2 is given by

$$\Delta v_2^- = \kappa_R \frac{2}{v_p} \Delta z \left. \frac{\partial V^{inc}}{\partial t} \right|_{t - \frac{z'}{v_p}} \tag{10.40}$$

An observer located at $z < z'$ would see an additional propagation delay $(z' - z)/v_p$, so the argument of Equation (10.40) would be modified to

$$t - \frac{z'}{v_p} - \frac{(z' - z)}{v_p} = t - \frac{(2z' - z)}{v_p} \tag{10.41}$$

and the signal at position z due to the crosstalk can be expressed as

$$\Delta v_2^- = \kappa_R \frac{2}{v_p} \Delta z \left. \frac{\partial V^{inc}}{\partial t} \right|_{t - \frac{(2z' - z)}{v_p}} \tag{10.42}$$

By superimposing the coupling between location z and the end of the line $(z = d)$, we obtain

$$v_2(z, t) = \kappa_R \frac{2}{v_p} \int_{z'=z}^{d} \left. \frac{\partial V^{inc}}{\partial t} \right|_{t - \frac{(2z' - z)}{v_p}} dz'$$

$$= \kappa_R \left\{ \left. V^{inc} \right|_{t - \frac{z}{v_p}} - \left. V^{inc} \right|_{t - \frac{(2d - z)}{v_p}} \right\} \tag{10.43}$$

Equations (10.39) and (10.43) describe the voltage induced into line 2 by the signal on line 1. We now check the assumptions made in the process of obtaining these expressions.

Consider the line 1 waveform consisting of the forward-going voltage and current waves

$$v_1(z, t) = f\left(t - \frac{z}{v_p} \right) \tag{10.44}$$

$$i_1(z, t) = \frac{1}{Z_0} f\left(t - \frac{z}{v_p} \right) \tag{10.45}$$

According to the reverse crosstalk equations, the signal on line 2 will contain negative-going voltage and current waves of the form

$$v_2(z, t) = \kappa_R \left\{ f\left(t - \frac{z}{v_p} \right) - f\left(t + \frac{z}{v_p} - \frac{2d}{v_p} \right) \right\} \tag{10.46}$$

$$i_2(z,t) = \frac{\kappa_R}{Z_0}\left\{ f\left(t - \frac{z}{v_p}\right) - f\left(t + \frac{z}{v_p} - \frac{2d}{v_p}\right)\right\} \qquad (10.47)$$

We wish to substitute this solution into the original transmission line equations for the coupled case, namely Equations (10.11) and (10.12). It is easily determined that

$$\frac{\partial v_2}{\partial z} = -\frac{\kappa_R}{v_p}\left\{ f'\left(t - \frac{z}{v_p}\right) + f'\left(t + \frac{z}{v_p} - \frac{2d}{v_p}\right)\right\} \qquad (10.48)$$

$$\frac{\partial i_2}{\partial z} = -\frac{\kappa_R}{Z_0}\left\{ f'\left(t - \frac{z}{v_p}\right) - f'\left(t + \frac{z}{v_p} - \frac{2d}{v_p}\right)\right\} \qquad (10.49)$$

$$\frac{\partial i_1}{\partial t} = \frac{1}{Z_0} f'\left(t - \frac{z}{v_p}\right) \qquad (10.50)$$

By substituting these expressions into

$$\frac{\partial v_2}{\partial z} = -L\frac{\partial i_2}{\partial t} - M\frac{\partial i_1}{\partial t} \qquad (10.51)$$

and balancing terms in the equation, we determine that the coefficients of the function $f'(t-z/v_p)$ combine to produce the equation

$$-\frac{\kappa_R}{v_p} = \frac{L\kappa_R}{Z_0} - \frac{M}{Z_0} \qquad (10.52)$$

Using

$$Z_0 = \sqrt{\frac{L}{C}} \qquad (10.53)$$

$$v_p = \frac{1}{\sqrt{LC}} \qquad (10.54)$$

and, from Equation (10.33),

$$\kappa_R = \frac{M}{2L} \qquad (10.55)$$

one sees that Equation (10.51) is satisfied. Similarly, the coefficients of $f'(t + z/v_p - 2d/v_p)$ combine to produce the equation

$$-\frac{\kappa_R}{v_p} = -\frac{L\kappa_R}{Z_0} \qquad (10.56)$$

which is immediately verified. Problem 10.8 asks the reader to check the validity of Equation (10.12) for reverse crosstalk by a similar process. For forward crosstalk, the equations are satisfied only because the ideal analysis concludes that $\kappa_F = 0$.

The preceding development is valid when the coupling between the lines is weak, which is usu-
ally taken to mean that $\Delta < 0.1$, $\kappa_R < 0.05$, and $\kappa_F < 0.05/v_p$. In the case of symmetric, homogeneous
lines, $\kappa_F = 0$. In practice, however, the idealized situation of homogeneous lines is seldom realized,
even if there is only one medium surrounding the trace, since nearby vias, traces crossing the path of
the line in question, etc., perturb the environment enough to cause some coupling. Therefore, we will
continue to assume that both κ_F and κ_R are likely to be nonzero. In the following, we will illustrate
the coupling process by example.

Figure 10.6 depicts two weakly coupled lines, with line 1 driven from the left end with a
voltage generator that produces a signal that ramps linearly to a constant voltage and remains at
that level. Both lines are matched at each end with their characteristic impedance to eliminate reflections. Both lines have a length $d = 10$ cm and a one-way transit time of $T = 500$ ps. Thus, the
propagation velocity on the lines is

$$v_p = \frac{d}{T} = \frac{10}{500} = 0.02 \text{ cm/ps} \tag{10.57}$$

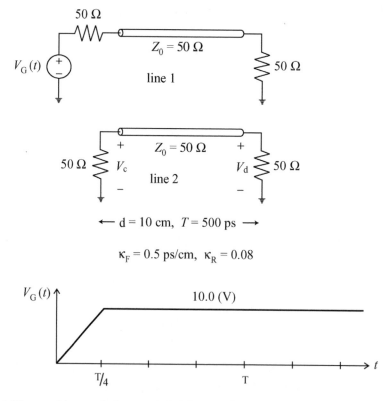

FIGURE 10.6: Two weakly coupled transmission lines and associated generator voltage.

The coupling coefficients are determined to be $\kappa_F = 0.5$ ps/cm and $\kappa_R = 0.08$. Our task is to determine the voltages observed across the loads at either end of line 2. The coupled signal results from the superposition of forward crosstalk, as described by Equation (10.39), and reverse crosstalk following Equation (10.43).

Figure 10.7 depicts the forward crosstalk induced at several locations on line 2 by the ramp function of Figure 10.6. Observe that the forward crosstalk signal is proportional to the derivative of the voltage waveform on line 1 and thus is localized in space and time. If the signal on line 1 exhibits a sharp rise time, the amplitude of the forward crosstalk signal will be large. Observe also that the signal grows in proportion with distance to the observer location down the line. Therefore, forward crosstalk is proportional to the length of the coupled-line system. If specialized to the end of line 2, Equation (10.39) yields

$$V_d = \kappa_F \left(\left. \frac{\partial V^{\mathrm{inc}}}{\partial t} \right|_{t - \frac{d}{v_p}} \right) d \tag{10.58}$$

Figure 10.7d shows V_d, the voltage across the load resistor at $z = d$.

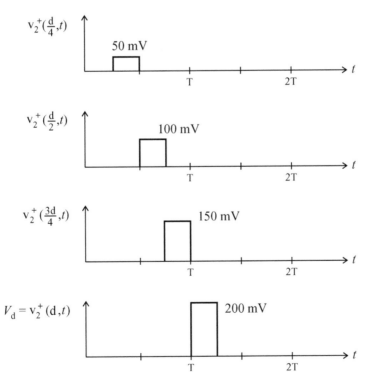

FIGURE 10.7: Forward crosstalk at locations on line 2 in the system of Figure 10.6.

Figure 10.8 depicts the reverse crosstalk observed at several locations on line 2. Observe that the reverse crosstalk near the $z = 0$ end of line 2 is of relatively broad duration, while that located near the $z = d$ end of line 2 is relatively narrow in time. Since the reverse crosstalk is proportional to the signal on line 1 and not its derivative, it tends to be smoother in character than the forward crosstalk. If specialized to the generator end of line 2, equation (10.43) becomes

$$V_c = \kappa_R \left\{ V^{inc}\big|_t - V^{inc}\big|_{t-\frac{2d}{v_p}} \right\} \tag{10.59}$$

Figure 10.8a shows V_c, the voltage across the resistor at the $z = 0$ end of line 2.

Now, suppose that the generator voltage of Figure 10.6 is changed to that of Figure 10.9, so that it has a finite duration in time and a relatively short rise time. Figures 10.10 and 10.11 depict the waveforms across the load resistors at the ends of line 2. The reverse crosstalk signal depicted in Figure 10.10 mimics the excitation in shape and duration, while the forward crosstalk signal in Figure 10.11 is of a sharper nature due to the derivative operator. In constructing these responses, it is helpful to first plot V^{inc}, V^{inc} delayed by time $2T$, and dV^{inc}/dt on line 1.

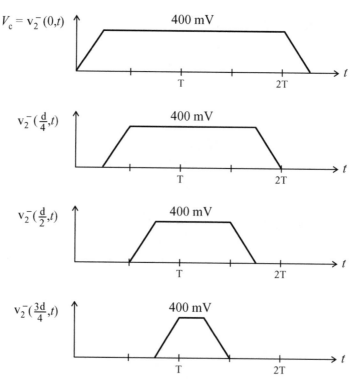

FIGURE 10.8: Reverse crosstalk at locations on line 2 in the system of Figure 10.6.

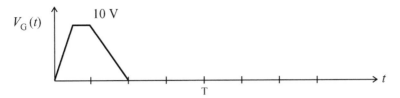

FIGURE 10.9: Alternate generator voltage to be used with the system in Figure 10.6.

Mismatches at the ends of either transmission line can be accounted for in the usual manner, by keeping track of the reflection coefficients and the associated forward- and reverse-going waves. A mismatch on line 1 will result in a reflected signal on line 1, contributing both "forward" and "reverse" crosstalk to either end of the second line. A mismatch on line 2 will result in the reflection of the appropriate induced signal on line 2.

Suppose we consider the transmission line system shown in Figure 10.12, where a short circuit load is placed at the far end of line 1. Consider the task of finding V_c, the voltage across the load resistor at $z = 0$, for the same generator signal and line parameters as in the previous example (Figure 10.9). In this situation, we must combine the reverse crosstalk induced by the forward-going wave on line 1 (already determined in Figure 10.10) and the forward crosstalk produced by the reflected signal on line 1. Although we use the "forward" crosstalk equation, the crosstalk is actually producing a reverse-going wave on line 2 since it is caused by the negative-going reflected waveform on line 1. Figure 10.13 depicts the forward crosstalk due to the reflection, obtained by conceptually considering the mirror image of the lines and incorporating the reflection coefficient $\Gamma_L = -1$ on line 1. These two results (Figures 10.10 and 10.13) are combined in Figure 10.14 to obtain the total voltage V_c.

An analogous process can be used to determine V_d, the voltage across the load resistor at $z = d$ on line 2. The forward crosstalk from the initial positive-going wave on line 1 (Figure 10.11) can

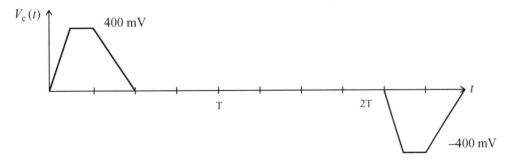

FIGURE 10.10: Forward crosstalk observed at the load end of line 2 due to the finite-duration generator voltage of Figure 10.9.

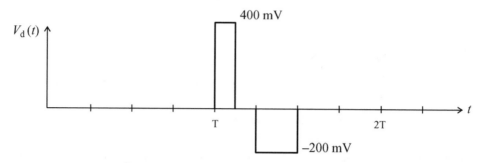

FIGURE 10.11: Reverse crosstalk observed across the resistor at the generator end of line 2, due to the alternative generator voltage of Figure 10.9.

be combined with the reverse crosstalk from the reflected signal on line 1 (Figure 10.15) to produce the total observed voltage V_d (Figure 10.16).

The preceding examples illustrate the crosstalk phenomenon. Additional examples of cross-talk calculations may be found in References [1–2]. As we have emphasized, the reverse crosstalk on line 2 mimics the shape of the forward-going waveform of line 1, while the forward crosstalk on line 2 is proportional to the derivative of the waveform on line 1. As a consequence, reverse cross-talk generally tends to be a smoother function than forward crosstalk, which often exhibits sharp spikes.

The preceding development is mathematical; the reader is likely to wonder if there is a more intuitive explanation of the difference between forward and reverse crosstalk. Fundamentally, it is rooted in a physical principle known as Faraday's law, which states that a changing magnetic field

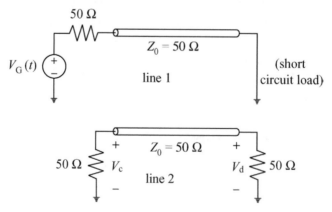

FIGURE 10.12: The system of Figure 10.6, with the matched load on line 1 replaced by a short circuit.

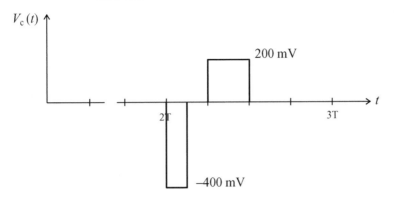

FIGURE 10.13: The forward crosstalk at the generator end of line 2, due to the reflected waveform on line 1.

induces a voltage in proportion to the rate of change (the time derivative) of the magnetic flux. This is why the forward-going signal on the second transmission line is proportional to the derivative of the signal on line 1.

While the identical mechanism is at work in exciting the reverse-going waveform on line 2, there is a unique additional consideration: as the waveform is excited on line 2, the resulting signal essentially reintegrates itself as it sweeps past the forward-traveling waveform on line 1. Consequently, the reverse crosstalk on line 2 is proportional to the incident waveform, and not to its derivative.

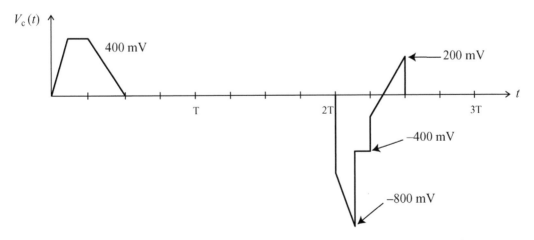

FIGURE 10.14: The voltage Vc produced at the generator end of line 2 in the system of Figure 10.12. This signal is the superposition of the waveforms in Figures 10.10 and 10.13.

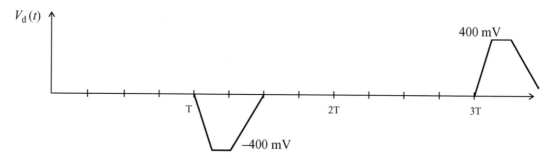

FIGURE 10.15: The reverse crosstalk at the load end of line 2, due to the reflected waveform on line 1.

From the standpoint of a circuit designer, crosstalk is undesired and should be reduced or eliminated if possible. As a rule, designers should attempt to avoid crowding signals together on any one part of a system. One way to minimize crosstalk is to design the layout to reduce the coupling coefficients between conductors as much as possible. This can be accomplished by increasing the spacing between traces, increasing the coupling to ground planes, or routing unused or grounded traces between the active lines. For particularly sensitive lines, a shield along both sides of the line may be fabricated from rows of metallized vias connected to or between ground planes. Grounded traces or a grounded via shield tend to contain the fringing fields and substantially reduce coupling to nearby lines. In addition, the length of lines running parallel to each other should be limited as much as possible. A complementary strategy is to limit the risetime and falltime of signals, to reduce the forward crosstalk on nearby lines, which is proportional to the time derivative of the signal.

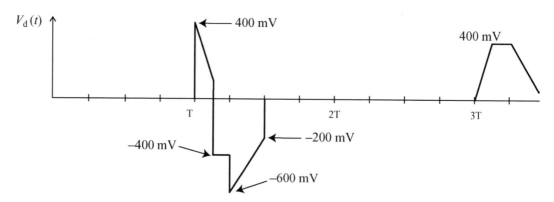

FIGURE 10.16: The voltage Vd produced at the load end of line 2 in the system of Figure 10.12. This signal is the superposition of the waveforms in Figures 10.11 and 10.15.

REFERENCES

[1]   S. Rosenstark, *Transmission Lines in Computer Engineering*. New York: McGraw-Hill, 1994.

[2]   S. S. Ang, "Electrical Design Considerations," in *Advanced Electronic Packaging: With Emphasis on Multichip Module*, ed. W. D. Brown, New York: IEEE Press, 1999.

PROBLEMS

10.1   *Fill in the blank:* When the coupling of two transmission lines causes signals to propagate on unintended pathways, the effect is called _____.

10.2   (a) *True or false:* One way to reduce the crosstalk between two lines is to place the lines further apart.

(b) *True or false:* The forward crosstalk signal on line 2 of a coupled system is proportional to the derivative of the positive-going incident waveform on line 1.

(c) *True or false:* The reverse crosstalk signal on line 2 of a coupled system is proportional to the derivative of the positive-going incident waveform on line 1.

10.3   The coupled line system shown below has a one-way transit time of $T = 500$ ps, line length $d = 10$ cm, and coupling coefficients $\kappa_F = -0.4$ ps/cm and $\kappa_R = 0.05$. The generator is excited with the given triangle waveform. Determine and plot the voltages V_c and V_d in the time interval $0 < t < 3T$, clearly labeling all voltage peaks.

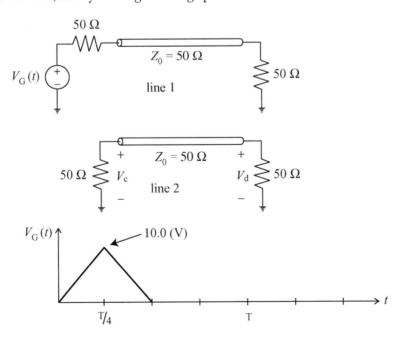

10.4 The coupled-line system shown in Problem 10.5 is used with different line parameters: the one-way transit time is $T = 15$ ns, the line length $d = 300$ cm, and the coupling coefficients are $\kappa_F = -0.2$ ps/cm and $\kappa_R = 0.04$. The generator is excited with the waveform shown below. Determine and plot the voltages V_c and V_d in the time interval $0 < t < 3T$, clearly labeling all voltage peaks.

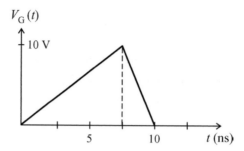

10.5 The figure below depicts two weakly coupled lines, with line 1 terminated at the right end in a short circuit load. The time delay on both lines is $T = 100$ ps and the line length is $d = 1.5$ cm. The coupling coefficients are $\kappa_F = 0.05$ ps/cm and $\kappa_R = 0.05$. The generator is excited with the trapezoidal waveform shown. Determine and plot the voltages V_c and V_d in the time interval $0 < t < 3T$, clearly labeling all voltage peaks.

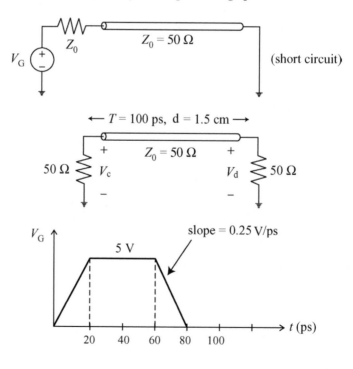

10.6   The weakly coupled system depicted below has line 1 terminated at the right end in a mismatched load. The propagation velocity on both lines is $v_p = 0.025$ cm/ps and the line length is $d = 2$ cm. The coupling coefficients are $\kappa_F = 0.02$ ps/cm and $\kappa_R = 0.06$.

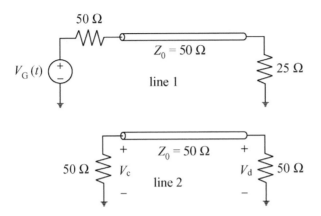

The generator is excited with the triangle waveform shown. Determine and plot the voltages V_c and V_d in the time interval $0 < t < 3T$, clearly labeling all voltage peaks.

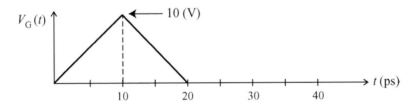

10.7   List five ways by which an electrical engineer can improve the design and layout of a high-frequency printed circuit board to minimize crosstalk.

10.8   For the forward-going waveforms given in Equations (10.44) and (10.45), substitute the approximate solution obtained for the signals on line 2 from (10.43) into equation (10.12), in order to demonstrate its validity for weakly coupled transmission lines.

Author Biography

Andrew F. Peterson received his B.S., M.S., and Ph.D. degrees in electrical engineering from the University of Illinois, Urbana–Champaign. Since 1989, he has been a member of the faculty of the School of Electrical and Computer Engineering at the Georgia Institute of Technology, where he is now Professor and Associate Chair for Faculty Development. He teaches electromagnetic field theory and computational electromagnetics and conducts research in the development of computational techniques for electromagnetic scattering, microwave devices, and electronic packaging applications. He is a fellow of the IEEE and a fellow of the Applied Computational Electromagnetics Society (ACES). He is also a recipient of the IEEE Third Millennium Medal.

Gregory D. Durgin joined the faculty of Georgia Tech's School of Electrical and Computer Engineering in fall 2003, where he teaches in the Electromagnetics group. He received his B.S. (1996), M.S. (1998), and Ph.D. (2000) degrees in electrical engineering from Virginia Polytechnic Institute and State University. In 2001, he was awarded the Japanese Society for the Promotion of Science (JSPS) Postdoctoral Fellowship and spent one year as a visiting researcher with Morinaga Laboratory at Osaka University. In 1998, he received the Stephen O. Rice prize (with coauthors Theodore S. Rappaport and Hao Xu) for best original journal article in the *IEEE Transactions on Communications*. Prof. Durgin also authored *Space–Time Wireless Channels*, the first textbook in the field of space–time channel modeling. He has received numerous teaching and research awards, published more than 50 technical articles, and serves as a frequent consultant to industry.

Printed in the United States
by Baker & Taylor Publisher Services